Uhler

OZONE IN WATER AND WASTEWATER TREATMENT

OZONE IN WATER AND WASTEWATER TREATMENT

Francis L. Evans III
Environmental Protection Agency
Cincinnati, Ohio
Editor

ann arbor science PUBLISHERS INC.
POST OFFICE BOX 1425 ● ANN ARBOR, MICHIGAN 48106

FOREWORD

The terms "environmental protection" and "ecology" are broadly accepted in our society. They are recognized by every stratum of industrial, commercial, and governmental organizations to be the driving force for an unprecedented unity of effort.

Since the impact of pollution is great, the efforts to reverse the trend must be greater. Much of the pollution of the nation's waterways comes from municipal and industrial sources which, fortunately, can be controlled. Water quality standards are becoming more strict, and the concept of minimal pollutional discharge is being approached.

For these reasons, we reevaluate the technology to satisfy today's needs in terms of current and future requirements for stream standards that preserve water quality, protect public health, and generally improve the environment.

While the "conventional" methods of waste treatment have been improved, standards of quality have surpassed the capabilities of many of these methods. Therefore, some old concepts which are no longer of value should be replaced by an advanced technology designed to meet future needs.

Ozonation is both an old and a new concept. It has been used for sixty-six years for sterilization, for not quite as many years in industrial and commercial applications, and most recently, for wastewater treatment.

In view of resurging interest in ozonation, it seems appropriate to put the subject in sharper focus through a series of articles on ozone applications in water and waste treatment technology. These presentations will show how an old concept can be reevaluated in terms of modern requirements.

The first chapter on the state of the art generally summarizes what the succeeding chapters present in detail. Early application in water disinfection, deodorizing, and decolorizing is reviewed. More recent application in water and wastewater treatment is discussed briefly.

One of the incongruous aspects of this chemical species is the difficulty encountered in the specific analytical determination. The lack of suitable techniques for analysis in low concentrations has been one of the obstacles limiting its study. This is the value of the second chapter which reviews the analytical methods for ozone in water and wastewater. The details of the development of a new analytical procedure as it applies to ozone disinfection are given along with a discussion of the data, a suggestion of the mechanism of reaction, and a description of the effects of interferences most likely to occur in water and wastewater analysis.

Current interest in ozone extends to its use as a powerful oxidant for organic matter in waste streams. Such a use was merely of academic interest until a few years ago, but this is not the case today. The third chapter explores the chemical considerations which held in the evaluation of the potential that ozone as an oxidant could have in wastewater treatment and pollution control applications. It presents a survey of the ozonation chemistry of organic compounds emphasizing functional groupings which react with ozone and the mechanism of these reactions in an aqueous medium.

With this information as a background, chapter four exemplifies how theoretical considerations hold up in the harsh reality of wastewater treatment situations. The current view of ozone treatment for further purification of secondary treatment plant effluents is expanded by the description of a pilot plant study.

Because of the interplay of many circumstances relating to the quality of environment, there has been considerable renewed interest in the use of ozone for disinfection. This potential use is in disinfection of heavily polluted waters used for municipal supplies and of wastewater treatment plant effluents. Although there has been much work done in pure and single-culture systems, chapter five points out that results from such studies are rarely comparable. Chapter six discusses the advantages and disadvantages of ozone as a disinfectant and describes a disinfection study using ozone in both pure-water systems and in a water containing various concentrations of raw domestic sewage.

Based upon promising results of apparent feasibility in proof of theory, the use of ozone and the practice of ozonation has been progressing from an art to a science. The last two chapters present supporting evidence of this fact. These two chapters also present practical aspects

of ozone and ozonation which demonstrate that ozone indeed
has advanced beyond the stage of being an academic curiosity.
The editor sincerely appreciates the time and effort each
contributor has devoted to this publication. Special thanks
is extended to Dr. Nina McClelland, National Sanitation
Foundation, for her special interest and help.

                              Francis L. Evans III

Cincinnati, Ohio
July, 1972

TABLE OF CONTENTS

CHAPTER I

OZONE TECHNOLOGY:

CURRENT STATUS

Francis L. Evans III

The literature is replete with such terms as "resistant," "perdurable," and "refractory" to describe materials that are not removed by waste treatment methods, whether these methods be conventional or "advanced."

The point is that, in view of today's needs and tomorrow's necessarily more stringent requirements, the quantity of materials remaining after treatment must be reduced. Although advances are being made to extend the potential of the conventional treatment methods, their potential has a limited value.

Chemical oxidation has the potential for removing from wastewaters those organic materials which are resistant to other treatment methods whether those methods are biological or the longer trains of processes known as tertiary (or advanced) treatment. Chemical oxidants have not been generally considered for use in domestic wastewater treatment because of the expense of materials and related costs and because of the lack of demand for such vigorous treatment.

However, it is not generally realized and accepted that, since the degree of treatment must be enhanced, there must be an increase in capital expenditure. It is likely, though, that the cost of this treatment will not be beyond the limits of economic acceptability.

The desirable characteristics of a chemical oxidant are that it be easily and economically available and that it not contribute secondary pollutants to the waste stream. The

oxidative destruction of organic compounds—resulting in
the formation of carbon dioxide, water, and oxygenated
fragments of the original molecule—occurs by bond fission.
According to Waters[1] this may be accomplished via two paths:
homolytic reactions, in which the electron pairs are dis-
rupted and one of the electrons is transferred to the
oxidant; and heterolytic reactions, in which the electron
pair is transferred or partially transferred as a unit to
the oxidizing substance. As a chemical oxidant, molecular
oxygen has no peer and, given sufficient time, will react
with any and all organic materials.

The possible reactions of molecular oxygen can be broadly
classified as oxygen insertion or noninsertion reactions.[2]
These are reactions in which oxygen itself combines with the
substrate, and reactions in which oxygen merely serves as
an oxidizing agent. Since the processes occur more or less
spontaneously, they are referred to as "autoxidations."[3]
However, the rates of reaction are generally too slow to be
of any value in waste treatment—biological systems excepted.
It is not improbable that autoxidation mechanisms proceed
via free radicals including chain mechanisms once the production of the free radical has been initiated. Autoxidation
is not, however, necessarily a chain reaction, but the
interpretation of some reactions can be made only on that
basis.[4]

The oxidative power of ozone has for some time been used
in the ozonolysis of olefins: the reaction of ozone with
a double bond which leads to cleavage of the bond. The
interest in ozonolysis reaction falls into three general
areas:[5] use of ozone to locate unsaturation in structure
determinations; use of the reaction synthetically (i.e.,
to convert unsaturation into ketones, aldehydes, alcohols,
or acids; and, finally, study of the mechanism of the
reaction.

The reactions of ozone as an oxidant appear to be of at
least two distinctive types.[6] One involves an electrophilic
attack by ozone; the other an ozone-initiated oxidation in
which ozone serves as the reaction initiator and oxygen is
the principal reactant. Examples of electrophilic attack
by ozone are the reactions of ozone with tertiary amines,
phosphines, arsines, sulfides, and sulfoxides. Decomposition of the substrate sometimes results. As an example,
consider the ozone oxidation of a dialkyl sulfide which
reacts according to the equation

$$R_2S + O_3 \longrightarrow R_2SO + O_2 \xrightarrow{O_3} R_2SO_2 + O_2 .$$

The well-controlled attack of ozone on the sulfur molecule
can best be explained[7] by the assumption that a terminal
oxygen of the ozone molecule executes an electrophilic
attack on the sulfur, forming a new bond with the sulfur.
The second and third atoms of oxygen are liberated as
molecular oxygen. Consideration of ozone oxidation,
therefore, should take into account not only oxidation via
ozonation mechanisms alone but also autoxidation mechanisms.
Examples of ozone-initiated reactions are the oxidative
reactions of ozone with aldehydes, ketones, alcohols,
ethers, and saturated hydrocarbon groupings. In these
reactions, ozone behaves as a radical reagent in an autoxidation process by mobilizing an additional number of oxygen
molecules. Consider the oxidation of benzaldehyde to
benzoic acid and perbenzoic acid by oxygen containing ozone.
Briner, *et al.*,[8] found a great deal of unreacted ozone, the
aldehyde appearing to be only slightly sensitive to the
oxidizing action of ozone. However, analysis of the aldehyde solution showed that it had been heavily oxidized and
that much more oxygen was taken up in this oxidation than
corresponded to the amount of ozone consumed. The reaction
steps represented by

$$R-\underset{\underset{O}{\|}}{C}-H + O_2 \longrightarrow R-\underset{\underset{O}{\diagup}}{\overset{\overset{O}{\diagdown}}{C}} - HO \longrightarrow R - \underset{\underset{O}{\|}}{C} - O - OH$$

$$\text{aldehyde} \qquad \text{peroxide} \qquad \text{peracid}$$

constitutes the actual autoxidation. Briner[9] has suggested
that the transformation of aldehyde to peracid proceeds by
a chain reaction in which certain short-lived radicals are
intermediates. Viewed from an energy standpoint, the reaction chains are energy-induced by ozone transformation
into oxygen, the energy released being that which is required
for the autoxidation reaction. While Long[10] states that
amino acids are not attached by ozone, Bergel and Bolz[11]
showed the reaction did occur and that aldehydes, ammonia,
and hydrogen peroxide are the major products.[12] The
oxidation of coal, carbon black, lampblack, and humic acids
by ozone has been reported.[13-15] The products were mainly
carbon dioxide and water-soluble acids.

WATER AND WASTEWATER APPLICATIONS

The events which chronicle the early history of ozone
and ozonation can be well described as meager. A duration

of about 125 years elapsed between the time that the pungent
odor of ozone was first noted by Van Marum in 1781, and
operation of the first major installation for sterilization
at Nice, France, in 1905. About 50 years elapsed (after
Van Marum) before ozone was recognized as indeed a unique
substance, and an additional 30 years passed before the
tri-atomic oxygen formula, O_3, was established in 1867.

The first use of ozone was for water sterilization
when, in 1893, Schneller, Vander Sleen, and Tindal constructed an industrial apparatus at Oudshoorn in Holland
for the sterilization of Rhine water after sedimentation
and filtration. The city of Paris, at the Saint Maur
filtration works, first used ozonation in 1898. Otto-ozone
equipment was initially used in 1907 at Nice in the first
major installation for the sterilization of a public water
supply. Ozonation might have become universal for disinfection in water treatment except for the introduction
in this country of cheap chlorine gas, a product of World
War I research in poison gases. In Europe and other places,
however, ozone for disinfection remained the method of
choice, and by 1936 there were more than 100 ozone installations in France and 30 to 40 elsewhere. Today, more than
500 municipalities in 50 countries use ozone as a
disinfectant.

On the other hand in this country—aside from the City
of Whiting, Indiana which used ozone in pretreatment for
removal of taste and odor from Lake Michigan water; and
Philadelphia, Pennsylvania, which used ozone to remove
taste and odor and manganese from Schuylkill River water—
there has been no major ozone installation for municipal
water or wastewater treatment, and chlorination remains
the principal method of disinfection.

Recently, however, concern has been expressed about
possible toxic effects resulting from the discharge of
chlorinated municipal and industrial wastes, which could
adversely affect aquatic life in a receiving stream. In
the search for alternatives to chlorine, ozone is being
considered. Although the literature on ozone disinfection
is plentiful, the work has generally been done in pure-
aqueous and single-culture systems. Despite these
simplifications, variations in experimental conditions
usually preclude a comparison of results, and such data,
having no relation to reality, cannot be extrapolated to
a practical situation.

As a result, the literature can be only generalized,
and can be done in this way. The use of ozone results in
disinfection but not necessarily sterilization. The action

of ozone is rapid. There appears to be a "threshold dose" which must be exceeded, prior to which there is slight bacterial kill and subsequent to which there is rapid kill. This threshold dose is a function of an "ozone demand" of the waste, which, in turn, is an inverse function of the degree of treatment prior to ozone application. Satisfying the ozone demand and disinfection are believed to occur simultaneously. Therefore, a greater degree of pretreatment is advantageous in two ways. One, the ozone demand is reduced by the removal of the more easily oxidizable substrate; and two, clumps or flocs which shield the organisms from the disinfectant are removed. The overall results are that a smaller dose of ozone is required to attain the threshold dose and a greater percent kill is attained with the smaller dose.

Although the disinfection of potable water is the largest application of ozone (aside from industrial uses), other uses of ozone have been made in the treatment of potable water. It is used for taste and odor control and for residual color reduction after conventional water purification by flocculation, sedimentation, and rapid sand filtration. Recently ozone has been applied more fundamentally, following discovery of its extremely effective action on organic colloids, in the "Micellization/ Demicellization" process and in the "Microzon" process (an abridgement of the M/D process). These processes result from an attempt to economize on the more "conventional" treatment processes. The intended result was to achieve a higher quality of water at a lower overall cost, and offset the higher capital cost of ozonation equipment when compared to the cost of eliminated processes and chlorination treatment. Laboratory and pilot plant studies have brought about the development of an ozonation/ coagulation/filtration process. The M/D Process is particularly suitable for surface waters containing organic matter that is often difficult to coagulate and requires uneconomical doses of electrolyte.

In contemporary uses of ozone, Boucher[16] has summarized the principal applications in sterilization of water, taste and odor control, iron and manganese removal and color reduction. Ozone has been used for odor control in sewage treatment plants in New York[17] and Florida.[18] The Midland (Michigan) Waste Treatment plant initially installed ozone equipment in order to evaluate its effectiveness in deodorizing the exhaust from the vacuum filtration system.[19] The trial was so successful in reducing sulfide odor that ozone is treating the exhaust

from the sludge storage tank building, and the remaining
open facilities (primary settlers, grit chamber and sludge
storage areas, and trickling filters) have been covered
with Styrofoam domes for exhaust gas ozonation. Similar
deodorization is being done at the Mamaroneck Sewage
Treatment Plant, Wards Island, and Owl's Head Sewage
Treatment facilities in New York.

In the United States, the renewed interest in ozonation
has been directed toward using not only the disinfection
powers of the chemical but also its oxidation potential.
Chemical oxidation as a unit process has not been generally
practiced in municipal wastewater treatment. However,
increasing demands on the water economy necessitate the
reduction of materials remaining after conventional waste
treatment. Although advances are being made which extend
the effectiveness of conventional treatment methods, the
limits of these methods are being approached, and new
technology must be devised.

Chemical oxidation may be applicable to many situations;
(1) as an alternative to carbon adsorption for use as a
tertiary treatment process (and when carbon adsorption and
regeneration cannot be economically justified); (2) as a
process to supplement carbon or synthetic resin adsorption;
(3) to extend the capabilities of intermediate treatment
plants; and (4) to enhance effluent quality in high-value
reuse situations. For these purposes, the most promising
oxidant class appears to be the active-oxygen species—
specifically ozone—which offers a vigorous oxidative
environment and rapid kinetics.

In evaluating the oxidative potential of ozone, there
have been more than a few laboratory studies of its effect
on certain specific constituents of wastewaters: anionic,
biologically-resistant, surfactants, all other types of
surfactants, saturated and unsaturated hydrocarbons,
chlorinated hydrocarbon insecticides, humic acids, iron,
manganese, cyanides, phenols—even anaerobic sewage microorganisms to increase their activity. As a result of these
laboratory studies, ozonation has been used to treat effluents from specialized industrial activities, and now is
being considered for use in the mixed-organics environment
of a municipal treatment plant.

Buescher and Ryckman[20] used ozone to reduce foaming of
ABS; and Evans and Rychman[21] reported 97% removal of ABS
from secondary sewage treatment plant effluent, accompanied
by an 80% reduction in the COD of the effluent. Ozonation
of ABS destroyed the biological inertness of the surfactant,
probably because the ozonides of the ABS solution served as

the oxidizable substrate for a mixed microbial population. Buescher *et al.* studied the destruction of lindane, aldrin, and dieldrin by various oxidants.[22] While ozone readily attacked lindane and aldrin and decreased the concentration of dieldrin, other oxidants as calcium hypochlorite, H_2O_2, Na_2O_2, and $KMnO_4$ (each at 40mg/liter dose) either had no measurable effect or only partial removal was achieved. Only aldrin was removed by hypochlorite and permanganate.

At present, there are no known full-scale ozone installations in a treatment plant for wastewater or effluent treatment, although recent studies might alter this. In 1966, a 2700-gallon-per-hour pilot plant was set up at the Eastern Sewage Works, Redbridge, London.[16,23] The pilot plant provided variable combinations of microstraining, prechlorination, ozonization, coagulation, and rapid sand filtration for the treatment of settled, biologically treated effluents. This M/D process employs the following three stages of treatment:

1. Microstraining for primary filtration,
2. Micellization/Demicellization to produce filterable microflocs, and
3. Rapid sand filtration for elimination of the microflocs.

Ozone is used in the second stage to bring about disruption of the colloidal state by attacking the hydrophillic grouping of organic macromolecular chains for the production of colloidal micelles. The production of "micellization" after ozonation is shown by the development of colloidal turbidity. The negatively-charged micelles require a dose of electrolyte to "demicellize" the water, and the resulting microflocs are subsequently sand-filtered. While the M/D process was in previous operation for water treatment at Roanne, France, and Constance, Germany, the Redbridge installation was the first attempt in waste effluent treatment. The process produced a final effluent low in suspended solids, clear and colorless, and the detergent content was reduced considerably. Concentrations of total solids, ammonia, and nitrate were virtually unaffected, but nitrite was oxidized. Dissolved organic matter was only slightly affected by the 20 mg/liter ozone dose. Ozone killed the vast majority of organisms present, including all the *Salmonella* and viruses. Chlorine produced lower counts than ozone; and chlorine followed by ozone was highly effective, all coliform and *E. Coli* counts being zero.

In a laboratory investigation supporting the Redbridge Pilot project, Gardiner and Montgomery[24] studied the effect of ozonation alone on the chemical composition of sewage effluents over a wide range of temperatures and ozone doses. The samples used were laboratory-settled trickling filter effluent; diluted, biologically treated milk waste; and an effluent from the treatment of detergent-free sewage. To some of the samples was added γ-BHC (Lindane) dieldrin, DDT, TDE; 10 mg/liter mixed phenols; 10 mg/liter Dobane JNQ (soft) or 10 mg/liter Dobane PT8 (hard) anionic detergent; and soft and hard non-ionic detergents. Some of the results of the study are as follows:

1. Ten minutes ozonation reduced organic carbon, measured by Beckman carbonaceous Analyzer, only slightly. One hour ozonation, on the other hand, effected a 27% reduction when 94 mg/liter ozone was absorbed. A greater reduction is achieved with a filtered and settled effluent than with a settled effluent.

2. Thirty to sixty minutes ozonation caused a reduction in the carbonaceous BOD; the samples further were likely saturated with dissolved oxygen.

3. The COD reduction in mg/liter was approximately equal to one-half the amount of ozone absorbed (in mg/liter). There was little difference in reduction between filtered and unfiltered settled samples.

4. Nitrite was oxidized rapidly to nitrate; organic nitrogen was slightly reduced; ammonia-N concentrations were unaffected.

5. Removals of phenol, pesticides, and detergents were in agreement with the removals found by other investigators.

While the former Redbridge pilot study demonstrated the usefulness of the M/D process or the Microzon process in the treatment of domestic waste treatment effluents and the associated laboratory study clarified the effect of ozone on some effluent constituents, the specific application of ozonation to the removal of residual organics from wastewaters was investigated by the Air Reduction Company.[25] The sample used was trickling filter effluent which was alum-clarified and sand-filtered to produce an influent with a reduced ozone demand. Both clarified and nonclarified secondary effluents were contacted with ozone for periods of time up to one hour. The influence of pH, method of gas dispersion, effluent pretreatment, and ozone concentration in the feed gas were evaluated in terms of ozone reactivity, reduction in the chemical oxygen demand, and reduction in total organic carbon content. This initial

study determined that the operation of a simulated six-
stage contactor reduced the COD from 32 to 13 mg/liter
and reduced the TOC from 12 to 9 mg/liter at a contact
time of one hour. Batch tests determined that the TOC
contains a fraction that is refractory to ozone oxidation,
a part of which can be removed by pre-clarification. It
was demonstrated that ozone efficiency for TOC and COD
reduction increased with increasing levels in the sample,
indicating that the more readily oxidizable organic compounds in the effluent consume ozone more readily. While
the COD showed a primary response to ozonation, no cleavage
of organic compounds is needed for a COD reduction. Only
on destruction of the organic residue, when oxidation
leads to formation of carbon dioxide, is there a reduction
in TOC. The fact that a TOC reduction was achieved demonstrates the oxidizing capabilities of ozone in the
destruction of organics refractory to biological treatment.
Cost estimates, based on the design of six-stage co-current
contacting system for 1-, 10-, and 100-mgd plants, assuming
80% ozone utilization, compared favorably with those for
activated carbon treatment plants.

Beyond controlled laboratory investigations and industrial waste treatment, pilot-scale ozonation plants have
been operated to attack the problem of pollution control
from a number of different angles. One plant was designed
to treat the combined sanitary and stormwater discharge in
Philadelphia (Fairmount Park), Pennsylvania, when storms
caused untreated tastes to overflow into the waterways.
In this instance, the combination of microstraining and
chlorine or ozone were compared. Another plant is
operating at the Hanover Park sewage treatment plant of
the Greater Chicago MSD. Here, an ozonator is treating
sewage effluent at 100 gpm to test ozone's effectiveness
in disinfection, color removal, and BOD reduction before
releasing the water into Lake Michigan. A two-stage injection system and a five-minute contact time is used.
When 0.1 ppm ozone concentration was maintained, it was
determined that the effluent was disinfected.

In the industrial waste treatment area, cyanide or
phenolic effluents from processes such as blast furnace
and open hearth operations, refining, and coke plants are
treatable with ozone, as are a variety of waste streams
from a synthetic polymer plant. Wastes resulting from
glycerol production which contain trace amounts of biologically inert or toxic materials have also been treated
with ozone. The Boeing plant at Wichita, Kansas, uses

more than 350 pounds of ozone per day to lower COD,
cyanide, phenols, oil, detergents, sulphides, and sulphites.
In another area, acid mine drainage has posed serious
problems in maintaining water quality, and available
methods of treatment have been less than satisfactory.
In evaluating ozone as an oxidizing agent, an ozone-
limestone system was found to offer the potential of
simplified process control, higher plant through-put,
removal of additional pollutants from AMD, and reduced
sludge-handling requirements, all at costs equal to or
less than those obtained using alternate techniques.

The removal of residual organic compounds by ozone is
currently receiving greater attention, and continued
effort should be expended in that area. Improvements in
classical ozone generation equipment is claimed by CEO,
one of the oldest manufacturers of ozone equipment.
Chromalloy American Corp. is claiming that its new Purogen
Activated Oxygen System overcomes ozone's two main draw-
backs: instability and high power costs. Among the
company's claim for Purogen are: (1) unlike other ozone
systems, it requires no cooling for product stabilization,
thus permitting a saving in power cost; (2) it produces
a sort of "activated oxygen," said to be even more active
than ozone; and (3) a 2 X 3 X 5-ft. unit can produce as
much ozone as a conventional ozone generator the size of
an automobile. Experimental Purogen applications with
industrial wastes have been tried, and the Los Angeles
Hyperion treatment plant is to try a Purogen unit on
municipal waste. The system has been used to control
algae and slime in water towers of air conditioning systems
in Los Angeles, Las Vegas, St. Louis, and Philadelphia at
a cost lower than required chemical costs alone. A new
system developed by Purification Sciences, Inc., includes
a low-cost ozone generation unit based on corona discharge
and a novel approach for mixing the ozone with the waste.

SUMMARY

This is briefly where ozone now stands as it has been
and is being applied in water and wastewater treatment.
Continuing laboratory and pilot plant studies will determine the treatability of waste streams from domestic,
municipal, and industrial sources. While much has been
accomplished in the past, the full potential has yet to
be developed. The task in doing this will not be easy
because, aside from mere demonstration of treatment

comparability or superiority on a technical scale, an image tarnished by misconception must be reestablished in proper perspective. Ozonation technology has not been dormant; and prior experiences with antiquated hardware, frequent breakdowns, and faulty and dangerous operation are just that—experiences of the past.

Contemporary ozonation hardware is the result of not only the modernization of conventional designs but also advanced conceptual design. New equipment makes use of modern high-temperature and acid-resistant materials, which have, in some instances, eliminated the necessity of incorporating the power- and space-consuming demands of some air-pretreatment and cooling-water devices. Advances in electronic design have provided new circuitry that offers advantages in power utilization and, of course, favorably affects production and therefore operating costs.

The realization of full benefits from ozone treatment will be attained through the cooperation of those advancing the science by scientific investigation, and those advancing the operational science. This cross-culturing could serve as a catalyst in making great strides to protect and improve the quality of life.

REFERENCES

1. Waters, W. A., *Mechanisms of Oxidation of Organic Compounds*. John Wiley and Sons, New York, 1964.
2. Martell, A. E., and Taquikhan, M. M., "Metal Ion Catalysis of Reactions of Molecular Oxygen." Presented at the Symposium on Inorganic Biochemistry, 151st National ACS meeting, April 1966.
3. Walling, C., *Free Radicals in Solution*. John Wiley and Sons, New York, 1957.
4. Uri, N., *Autoxidation and Antioxidants*. W. O. Lundberg, ed., Interscience, 1961.
5. Murray, R. W., "The Mechanism of Ozonolysis." *Accts. Chem. Res.*, *1*:10, 313 (1968).
6. Bailey, P. S., "The Reactions of Ozone with Organic Compounds." *Chem. Rev.*, *58*, 925 (1958).
7. Maggiolo, A., and Blair, E. A., "Ozone Oxidation of Sulfides and Sulfoxides." *Ozone Chemistry and Technology*, Advances in Chemistry Series #21, American Chemical Society, 1959.
8. Briner, E., Demolis, A., and Paillard, H., "The Ozonization of Aldehydes. The Action of Ozone and the Participation of Oxygen in the Reaction." *Helv. Chim Acta*, *14*, 794 (1931); Chem Abstr., *25*:4867 (1931).

9. Briner, E., "Accelerating Action of Ozone in the Autoxidation Process." Ozone Chemistry and Technology, Advances in Chemistry Series #21, American Chemical Society, 1959.

10. Long, L., "The Ozonization Reaction." Chem. Rev., 27, 437 (1940).

11. Bergel, F., and Bolz, K., "Uber Die Autoxydation von Aminosauidervaten und ihr Ablau durch Ozon." Hoppe-Seyler's Z. Physiol. Chem., 220, 20 (1933).

12. Schönberg, A., and Moubacher, R., "The Strecker Degradation of α-Amino Acids." Chem. Rev., 50, 261 (1952).

13. Ahmed, M. D., and Kinney, C. R., "Ozonization of Humic Acids Prepared from Oxidized Bituminous Coal." J. Amer. Chem. Soc., 72, 559 (1950).

14. Brooks, V. T., Lawson, G. J., Ward, S. G., and Dobinson, F., "Chemical Constitution of Coal." Fuel, 35:4, 385 (Oct. 1956).

15. Kinney, C. R., and Friedman, L. D., "Ozonization Studies of Coal Constitution." J. Amer. Chem. Soc., 74, 57 (1952).

16. Boucher, P. L., "Microstraining and Ozonation of Water and Wastewater." Presented at the 22d Annual Industrial Waste Conference, Purdue University, 1967.

17. Griffin, G. E., "Good Neighbor Plant." American City, 80, 99 (1965).

18. Miller, F. J., "Upline Sewage Treatment." Water and Wastes (Eng.), 3:12, 52 (1966).

19. Maass, A. E., "Successful Odor Control at Michigan Waste Water Treatment Plant." Water and Sewage Works, 114, 322 (1967).

20. Buescher, C. A., and Ryckman, D. W., "Reduction of Foaming of ABS by Ozonation." Proceedings of the 16th Industrial Waste Conference, Purdue University, 1961.

21. Evans, F. L. III, and Ryckman, D. W., "Ozonation Treatment of Wastes Containing ABS," Proceedings of the 18th Industrial Waste Conference, Purdue University, 1963.

22. Buescher, C. A., Dougherty, J. H., and Skrinde, R. T. "Chemical Oxidation of Selected Organic Pesticides." J. Wat. Poll. Cont. Fed., 36 (8), 1005 (1964).

23. Boucher, P. L., Lowndes, M. R., Truesdale, G. A., Mann, H. T., Windle Taylor, E., Burman, N. P., and Poynter, S. P., "Use of Ozone in Reclamation of Water from Sewage Effluent." Instn. Publ. Hlth. Engrs. J., 67, 75 (1968).

24. Gardiner, D. K., and Montgomery, H. A. C., "The Treatment of Sewage Effluents with Ozone." Water and Waste Treatment, 12:3, 92 (1968).

25. Halfon, A., Huibers, D. Th. A., and McNabney, R.,
    "Organic Residue Removal from Waste Waters by Oxidation
    with Ozone." Presented at the Division of Water, Air,
    and Waste Chemistry, American Chemical Society National
    Meeting, Atlantic City, September 1968.

CHAPTER II

ANALYTICAL METHODS FOR OZONE

IN WATER AND WASTEWATER APPLICATIONS

R. F. Layton

Ozone, O_3, an allotropic form of oxygen discovered by Schonbein in 1840, has extremely important significance in various environmental health areas including air pollution, water pollution, and industrial hygiene studies; and the analytical determinations of this material are of singular importance.

OZONE IN AIR POLLUTION STUDIES

Presently ozone has importance in the reaction sequences describing air pollution. It is a major constituent in photochemical smog and the major (up to 90%) component of the photochemical oxidant, defined as substances that oxidize a selected reagent not oxidizable by oxygen. Ozone has been shown to be formed during photochemical oxidation of hydrocarbons in the presence of nitrogen dioxide.[1]
The diurnal fluctuations of oxidant and ozone was one of the earliest characteristics established in the study of photochemical smog. In the predawn hours, there is an increase in carbon monoxide, nitric oxide, and hydrocarbons, while ozone and nitrogen dioxide remain negligible due to the absence of photochemical reactions. It is in the early morning hours after dawn that nitric oxide and hydrocarbons are at a maximum concentration. However, nitrogen dioxide begins to be generated at a substantial rate in one to two hours. Nitric oxide is reduced to low levels as it converts

15

to nitrogen dioxide. The nitrogen dioxide reaches a peak, and with the disappearance of nitric oxide comes the first appearance of ozone. The latter begins to accumulate as the photochemical oxidation proceeds. Nitrogen dioxide decreases as the ozone concentration increases. Ozone reaches a maximum concentration during early afternoon, then gradually declines over the next several hours. As the afternoon automobile traffic increases, the nitric oxide concentration increases, scavenging the remaining traces of ozone by early evening. Ozone concentration during darkness rarely exceeds 0.05 ppm.[1]

The background atmospheric concentration of ozone in surface air at sea level is approximately 0.01 to 0.03 ppm,[2] but, during a severe smog day in the Los Angeles area, the ozone concentration often reaches 0.5 ppm. The maximum content detected as an urban contaminant in outdoor air in Los Angeles was 0.99 ppm during a smog episode in 1956.[1]

Effects of Ozone on Humans

The primary concern over the increases in atmospheric ozone concentrations is its effect on the health of man. The clinical effects immediately recognized from inhaling ozone range from dislike of odor; headaches; dryness of the mucous membranes of the mouth, nose, and throat; and changes in visual acuity. Continued exposure results in more serious changes such as functional derangements of the lung, pulmonary congestion, and edema. Because human populations vary widely in their response (susceptibility) to ozone, for each of these manifestations a threshold range can be assigned.[1] For odor detection, 0.02 to 0.05 ppm is usually accepted for the general population. Irritation of the nose and throat in sensitive people has been known to occur at just the level of detection (0.05 ppm) by others,[3] dryness of the upper respiratory mucosa at 0.1 ppm, and dryness of the throat begins at levels slightly above 0.1 ppm. Increasing the levels to approaching 1.0 ppm for periods longer than 30 minutes produces headaches.[4] Ozone is not felt to produce eye irritation at these levels.[1] Changes in visual parameters such as visual acuity and extraocular muscle balance have been observed at levels from 0.2 to 0.5 ppm,[5] while changes in pulmonary functions such as changes in vital capacity occur when levels range between 0.1 and 1.0 ppm.[1]

OZONE IN WATER AND WASTEWATER TREATMENT

Ozone has found limited but increasing application in water and wastewater technology. First applied to the sterilization of water by Schonbein at Metz in 1840,[6] several recent research efforts have been reported. Studies have shown ozone to be effective as a bactericide and capable of removing color, taste, and odor from various waste effluents.

Ozone appears to offer several distinct advantages when used as a wastewater disinfectant:

1. The oxygen released when ozone is reduced can act as a means of insuring dissolved oxygen levels in the wastes being treated, thus reducing the nuisance potential of the waste being discharged and reducing both the extent and duration of the "oxygen sag" in the receiving stream.

2. Is said to eliminate many taste and odor problems often encountered when chlorination procedures are applied to phenolic and other wastes.

3. Ozone eliminates the possible formation of potentially dangerous end-products of halogenation procedures which could be similar in structure to pesticides.

4. While ozone has been used as a disinfectant, at the same time it possesses a high oxidation potential. An added advantage to disinfection of wastes by this technique results in the reduction of both the BOD and COD load to the receiving water, and will effectively reduce taste, odor and color.

5. As applied in water treatment technology, ozone is said to sterilize water rather than merely disinfect it at concentrations ordinarily used for chlorination procedures.

INTERRELATIONSHIPS OF HEALTH-AIR-WATER

To apply ozone to water and wastewater systems, the plant operator, technician, or researcher using this gas must be aware of the potential problems involved.

First, ozone is toxic and dangerous to humans, plant life, and aquatic forms. Although a half-life of 15 minutes has been reported for aqueous systems, ozone can exist in air for long periods of time. Plant life around treatment works using ozone has been destroyed in past research projects and interested workers must remain cognizant of the dangers when using this material.

To solve one environmental health problem such as reduction of BOD while simply creating air pollution or industrial
health problems must not be allowed when working with ozone.
In the same context, it is as important to monitor analytically
the atmospheric ozone levels around the treatment works as
the dosage and residual of ozone in the water and wastewater
systems.

CHEMISTRY OF OZONE

Ozone, a colorless gas at room temperature, is an allotropic form of oxygen, one and one-half times as dense,
with a molecular formula O_3 and molecular weight of 48.00
grams per mole. Ozone has a peculiar odor similar to that
of chlorine and this characteristic pungent odor has often
been detected in the atmosphere after electrical storms
and around other electrical discharges.

Like oxygen, ozone is a supporter of combustion and is
a powerful oxidizing agent, attacking almost all organic
compounds. Ozone which is one of the most reactive gases
known--in fact, the fourth most powerful oxidizing agent
known (only F_2, F_2O, and $O\cdot$ are better) is thought to have
a mechanism of oxidation related to the following reaction:

$$O_3 \rightarrow O_2 + O\cdot$$

where nascent oxygen produces a high-energy oxidation via
a free radical reaction.

GENERATION OF OZONE

From a few localities in this country and from numerous
cities in England and France, data concerning treatment of
water with ozone have been published. In general, the cost
of ozonation of water has been high, one of the reasons
being the unavailability of low cost generators of this gas.
Ozone can be produced by three techniques: (1) electrical
discharge, (2) electrolysis of perchloric acid, and
(3) ultraviolet exposure of oxygen. The first two of these
procedures give high concentrations of ozone, while the
ultraviolet lamps produce approximately .003 g ozone/hour
for each 1/100 of a watt. The only practical method of
large-scale ozone production has used the electrical
discharge principle.

On a laboratory scale, the generation of ozone for
analytical methods development in both air and water systems

can be accomplished in two ways. Figure 1 illustrates
laboratory equipment involving the use of an ultraviolet
light to generate an ozone stream for collection in an
impinger flask.

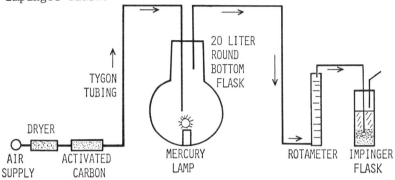

Figure 1. Laboratory Apparatus for Generation of Ozone
Using Ultraviolet Light

Figure 2 shows the use of a commercial ozone generator
attached to an 18-foot lucite contact column, again used
for preparation of ozone solutions. Both techniques are

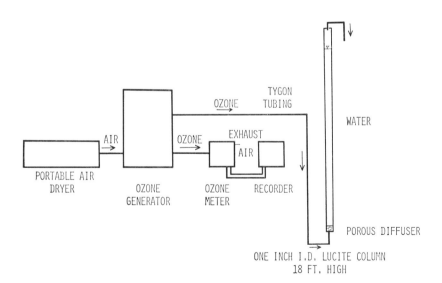

Figure 2. Experimental Apparatus for Generation of Ozone
Used at Mill Creek Sewage Treatment Plant

useful in laboratory and pilot scale work with the electrical
ozone generator representing a more expensive but more
dependable system.

ANALYTICAL METHODS FOR OZONE IN AIR

The analysis of ozone in air at sub-ppm levels has presented a difficult challenge for chemists for many years.
The oxidation of KI to iodine was the first analytical method
for determining ozone (or more correctly total oxidant) in
air and various modifications at alkaline, acid, and neutral
conditions have been developed. The appearance of new methods
for ozone in air have illustrated the limitations of existing
methodology. Literature involving wet chemical procedures
such as sodium diphenyl amine sulfonate[7] and 1,2-di-(4-pyridyl)
ethylene and others[8-10] has been presented in recent years.
Various instrumental procedures including ultraviolet,[11]
infrared,[12] and galvanic[13] have also been presented. Table
1 lists these techniques. Although much study has been
devoted to these methods, additional research is warranted
and is now in progress.

ANALYTICAL METHODS FOR OZONE IN WATER

The analysis of ozone in water at ppm levels has represented one of the limitations to conducting basic research
on the applicability of ozonation to water and wastewater
treatment. The instability of ozone in water has often
been discussed,[14] and most analytical methods for this gas
involve low pH solutions and temperatures near the freezing
point of water. The most commonly used method for ozone in
water involves the oxidation of potassium iodide, usually
in acid solution, where the following reaction is involved:

$$O_3 + 2H^+ + 2I^- \rightleftarrows O_2 + I_2 + H_2O$$

Serious limitations have been experienced by persons using
the iodometric methods, including non-stoichiometric iodine
formation under both neutral and alkaline conditions.[15]
Other studies have contraindicated high iodine production
in acidic solutions.[16]

Other methods presented have involved the oxidation of
manganous and ferrous ions.[17] While various instrumental
methods are also commonly employed, most involve monitoring
the oxidation of iodide ion to iodine. A list of the common

Table 1. Analytical Methods for Oxidants (Ozone) in Air

Method	Utilization of	Interferences or Limitations	Refs.
Potassium iodide; alkaline, acid, and neutral conditions	$2KI \rightarrow I_2$; titrate with reducing agent	Any oxidant interferes, NO_2 and SO_2 notably	18
1,2- di-(4-pyridyl) ethylene	Ozonide formation which cleaves forming pyridine-4 - aldehyde	Method applicable at temperatures >16° C only alkenes, SO_2, and NO_2 interfere; perhaps measures total oxidant	8, 9
Sodium diphenyl-aminesulfonate	Undefined color formed, measure at 593 mµ	NO_2 interferes	7
Phenolphthalein oxidation	Oxidation of colored indicator	Any oxidant interferes	19
Nitrogen dioxide equivalent method			10
Long path ultra-violet; long path infrared; galvanic monitoring cell	Instrumental	Lost sensitivity and costly instrumentation	11 12 13
Layton & Quick method	Aromatic isocyanate oxidation	Will detect only labile-containing peroxy compound	20 21

methods for ozone in water is presented in Table 2. A new
technique has been recently presented which will be discussed
in detail.

Table 2. Analytical Methods for Oxidants (Ozone) in Water

Method	Utilization of	Interferences or Limitations	Refs.
Potassium iodide; alkaline, acid, and neutral conditions	$2KI \rightarrow I_2$	Most oxidants interfere (oxygen interferes)	18
Ferrous ion oxidation (Luther & Inglis)	$Fe^{++} \rightarrow Fe^{+++}$	Results felt to be low	15
Manganese oxidation and orthotolidine	$Mn^{++} \rightarrow Mn^{+++}$		17
Visible region spectrophotometry	Molar absorptivity of 2500 to 3000 at 260 mu	Detection limits for 1 cm cell $10^{-3}M$	14
Oxidation of leuco crystal violet (Layton & Kinman)	Redox indicator; leuco crystal violet	Newest technique (under investigation)	22
Various instrumental methods using KI oxidation	$2KI \rightarrow I_2$	Same as KI above	--

LEUCO CRYSTAL VIOLET FOR OZONE DETECTION

While performing disinfection studies with ozone, the
need for a new sensitive method for ozone became apparent.
Leuco crystal violet, a salt of crystal violet A, was first
used by Black and Whittle[23] for monitoring both chlorine and
iodine residuals in swimming pools.
The application of this compound to monitor ozone in both
air and water has been recently reported by Layton and

Kinman.[22] This technique involves a colorimetric procedure utilizing a wavelength of 592 mμ in acidic solution. A molar absorptivity of 1.5 X 10^5 liters/cm mole has been calculated and assuming an absorbance of 0.1 in a 10-cm sample cell, sub-ppm levels can be readily detected. The reaction involved would appear to be as follows:

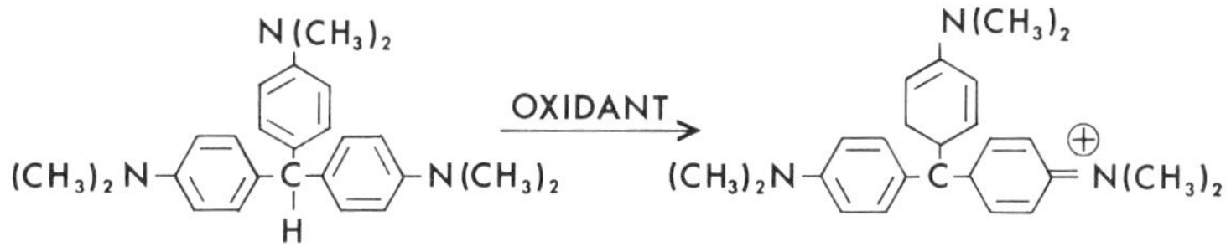

Leuco crystal violet   Crystal violet

where the ozone color observed is more similar to the purple color obtained with iodine than the blue color with chlorine.

The hydride transfer mechanism suggested for the oxidation of leuco crystal violet by Black and Whittle[23] could occur if one considers the reasonance structures of ozone. Ozone is a bent molecule which can be written as:

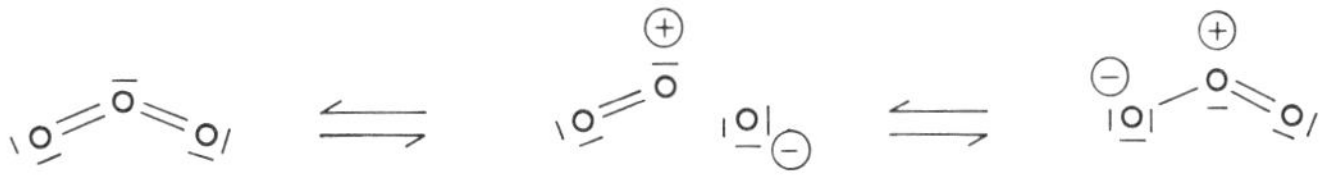

where the hydride transfer mechanism would actually occur via the following scheme:

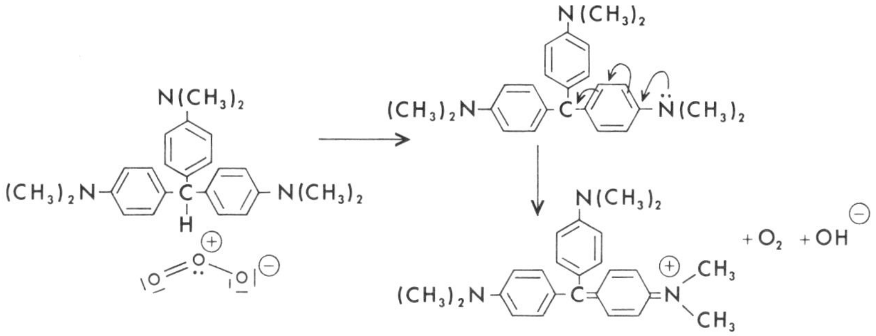

Another experimental observation supporting this hypothesis is the fact that hydrogen peroxide did not oxidize the leuco crystal violet at pH values 4, 7, 10, under the test conditions studied.

The color was found to be stable for an indefinite period of time (actually monitored for 44 days) indicating the possibility of using the color in preparation of a test kit for the monitoring of ozone, as is currently done with I_2 and Cl_2.

CALIBRATION CURVE FOR OZONE

Using the equipment previously described in Figure 1, ozone was generated into a collection flask for various periods of time, using constant air volumes. The data obtained are shown in Figure 3 and their adherence to a linear Beer's law plot is indicative of the utility of the method.

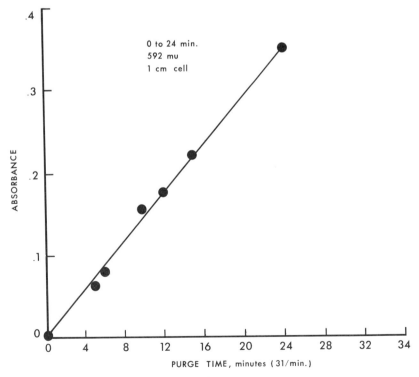

Figure 3. Calibration Curve for Ozone in Air

Standardization of the ozone/air stream produced by the experimental equipment illustrated with KI and thiosulfate techniques indicate that sub-ppm ozone solutions can be analyzed using this procedure. By varying the purge time and/or volume of air sampled, one can increase the sensitivity to the desired level. It is at least 10 times as sensitive as the standard KI procedures. The molar absorptivity values calculated for the color produced for iodine and chlorine are 1.90×10^5 and 1.42×10^5 respectively. Assuming ozone forms the same leuco crystal violet, it is reasonable to assume that about 10^{-8} $\underline{M}$ solutions of ozone would represent a minimum detectable amount (assuming a 10-cm cell and absorbance values of .1).

GENERATION OF OZONE-WATER SAMPLES

Figure 2 shows the experimental equipment used to generate ozone at the Mill Creek Sewage Treatment Plant, Cincinnati, Ohio. Table 3 represents absorbance values obtained with the leuco crystal violet method plotted vs. ppm ozone as determined by the KI procedure outline in Standard Methods.[24] These data, when plotted as Figure 4 gave a smooth, linear plot through the origin representing adherence to Beer's Law.

Table 3. Calibration Data for Ozone in Water (0 to 2.5 ppm)

Absorbance at 592 mµ	ppm Ozone by KI
.003	0
.015	0.09
.050	0.18
.090	0.46
.170	0.92
.252	1.38
.335	1.84
.390	2.07
.429	2.30
.430	2.36

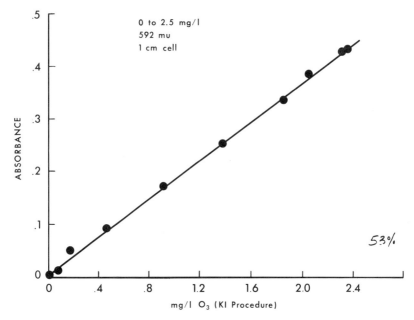

Figure 4. Calibration Curve for Ozone

SUMMARY

Although ozone was first applied to water in 1840, the utilization of this material has not developed substantially until recent years. One limiting factor in working with ozone systems has been the need for a specific and sensitive analytical technique to monitor this compound in aqueous systems.

Analytical methods for ozone have increased in number in recent years due primarily to its importance in air pollution problems concerned with photochemical smob. The application of leuco crystal violet to ozone determinations is the newest method presented to date and appears to be a promising method, although further research is needed.

Research to develop new methods capable of detecting sub-ppm levels of ozone in highly contaminated waters is much needed. With an increasing emphasis on ozonation of wastes, such technology is a necessity.

BIBLIOGRAPHY

1. Stern, A. C., Air Pollution. Vol. I, 2d Ed. Academic Press, New York (1968).
2. Junge, C. E., Air Chemistry and Radioactivity. Academic Press, New York (1963).
3. Henschler, A., Stier, H., Beck, H., and Newmann, W., "Olfactory Threshold of Some Important Gases and Manifestations in Man by Low Concentration." Arch. Gewerkepathol. Gewerkehyg, *17*, 547 (1960).
4. Wilsha, S., "Ozone: Its Physiological Effects and Analytical Determination in Laboratory Air." Acta Chem. Scand., *5*, 359 (1951).
5. Lagerwerff, J. M., "Prolonged O_3 Inhalation and Its Effects on Visual Parameters." Aerospace Med., *34*, 479 (1963).
6. AWWA Manual, Water Quality and Treatment, 2d ed., (1950).
7. Bovee, H. H., and Robinson, R. J., "Sodium Diphenylaminesulfonate as an Analytical Reagent for Ozone." Anal. Chem., *33*:8 (1961).
8. Hauser, T. R., and Bradley, D. W., "Specific Spectrophotometric Determination of Ozone in the Atmosphere Using 1, 2-Di-(4-Pyridyl) Ethylene." Anal. Chem., *38*: 11, 1529 (1966).
9. Hauser, T. R., and Bradley, D. W., "Effect of Interfering Substances and Prolonged Sampling on the 1, 2-Di-(4-Pyridyl) Ethylene Method for Determination of Ozone in Air." Anal. Chem., *39*:10, 1185 (1967).
10. Saltzman, B. E., and Gilbert, N., "Microdetermination of Ozone in Smog Mixtures: Nitrogen Dioxide Equivalent Method." Am. Ind. Hyg. Assoc. J., *20*, 379 (1959).
11. Renzetti, N. A., "Automatic Long-Path Ultraviolet Spectrometer for Determination of Ozone in the Atmosphere." Anal. Chem., *29*:6, 869 (1957).
12. Hanst, P. L., et al., "Absorptivities for the Infrared Determination of Trace Amounts of Ozone." Anal. Chem., *33*:8 (1961).
13. Hersch, P., and Deuringer, R., "Galvanic Monitoring of Ozone in Air." Anal. Chem., *34*:7 (1963).
14. Alder, M. G., and Hill, G. R., "The Kinetics and Mechanism of Hydroxide Ion Catalyzed Ozone Decomposition in Aqueous Solution," J. Amer. Chem. Soc., *72*, 1884 (1950).
15. Ozone Chemistry and Technology, American Chemical Society, Advances in Chemistry, Washington, D.C. (1959).
16. Boelter, E. D., Putnam, G. L., and Lash, E. I., "Iodometric Determination of Ozone of High Concentration." Anal. Chem., *22*:12, 1533 (1950).

17. Zehender, F., and Stumm, W., <u>Mitt</u>. <u>Gebiete</u> <u>Lebensm</u> <u>U</u>. <u>Hyg</u>., *44*, 206 (1953).
18. Selected Methods for the Measurement of Air Pollutants, U.S. Dept. of HEW Publication No. 999-AP-11.
19. Haagen-Smit, A. J., and Brunelle, M. J., <u>J</u>. <u>Air</u> <u>Pollution</u>, *1*, 51 (1958).
20. Quick, Q. Q., and Layton, R. F., "Analytical Significance of the Aromatic Isocyanate/Peroxy Reaction," <u>Anal</u>. <u>Chem</u>., *40:7*, 1158 (1968).
21. Layton, R. F., and Knecht, L. A., "Further Investigation of the Colorimetric Reaction between Aromatic Isocyanates and Peroxy Compounds," <u>Anal</u>. <u>Chem</u>., *43*, 794 (1971).
22. Layton, R. F., and Kinman, R. N., <u>A</u> <u>New</u> <u>Method</u> <u>for</u> <u>Determining</u> <u>Ozone</u>. Presented at the National Specialty Conference on Disinfection, University of Massachusetts, July 15, 1970.
23. Black, A. P., and Whittle, G. P., "New Methods for the Colorimetric Determination of Halogen Residuals. Part I: Iodine, Iodide and Iodate," <u>J</u>. <u>AWWA</u>, *59:4* (1967).
24. <u>Standard</u> <u>Methods</u> <u>for</u> <u>the</u> <u>Examination</u> <u>of</u> <u>Water</u> <u>and</u> <u>Waste-</u> <u>water</u>, 12th ed., APHA, AWWA, WPCF (1965).

CHAPTER III

ORGANIC GROUPINGS REACTIVE TOWARD OZONE

MECHANISMS IN AQUEOUS MEDIA

Philip S. Bailey

The other chapters of this book deal largely with practical applications, analytical procedures, and engineering
problems concerned with the use of ozone in water purification and sewage and chemical waste treatment. The materials
with which ozone reacts in such treatment are mostly organic
substances (compounds of carbon). This includes even the
living organisms toward which disinfection is directed.
Thus, a brief survey of the organic chemistry of ozone
seems most appropriate and is the purpose of this chapter.
Attention will be called to the ozone molecule and its
great and versatile oxidizing ability in regard to organic
substances. A brief survey of present-day concepts of the
mechanisms of these reactions will be made, with emphasis
placed on ozonations in aqueous media whenever possible.
Key references will be given rather than an exhaustive
listing of the literature.

Unfortunately, very few detailed studies have been made
into the chemical pathways involved in ozonation of organic
substances in water. In most studies in aqueous media,
either a cosolvent or a water emulsion or suspension has
been employed. Since in many cases there should be no
appreciable difference in the chemistry involved, for this
survey if no studies of the reactions of certain organic
groupings have been made in aqueous media, the reactions
in nonaqueous media will be discussed. No attempt will be
made to evaluate the problems associated with the practical
applications of these reactions to water treatment. It is

hoped, however, that the following discussion will provide
a useful theoretical background for those concerned with
such problems.

THE OZONE MOLECULE

A study of the microwave spectrum of the ozone molecule
has shown it to be nonparamagnetic, to have an obtuse angle
of 116°49' and equal oxygen-oxygen bonds of 1.278Å length,
and to possess a very low dipole moment, 0.49-0.58 debye.[1]
On this basis, it can be described as a resonance hybrid
of the canonical forms (I) shown in Figure 5. A simple
molecular orbital picture of the molecule is described by
II in Figure 5, in which each oxygen atom is sp^2, the sp^2
orbitals are filled, either through bonding or with un-
shared electrons, and overlap of the p orbitals provides
π molecular orbitals for the 4 π electrons involved.[1,2]

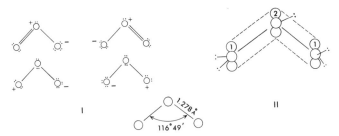

Figure 5. The Ozone Molecule

These descriptions of the ozone molecule lead to the
expectation that it should behave as a 1,3-dipole,[3,4] an
electrophile, or a nucleophile.[5] There is no reason to
suggest that ozone should behave as a radical, since it is
diamagnetic,[1] or that decomposition to molecular and atomic
oxygen should play a significant role in its reactions with
liquid solutions or suspensions in liquids. Such reactions
are generally carried out at "room temperature" or below,
and at such temperatures the decomposition rate would be
much slower[6] than the rates of the reaction types which
will be discussed herein. Possible exceptions to this
involve higher temperatures and the use of catalysts and/or
photolysis.[7]

Because ozone has an electronegative oxidation potential
exceeded only by that of fluorine,[8] it is a very versatile

and powerful oxidizing agent, reacting with other molecules
in a variety of ways. Among the organic groupings which
can be oxidized by ozone are olefinic and acetylenic
carbon-carbon double and triple bonds; aromatic, carbocyclic and heterocyclic molecules; carbon-nitrogen and
similar unsaturated groupings; nucleophilic molecules such
as amines, sulfides, sulfoxides, phosphines, phosphites,
arsines, selenides, etc.; carbon-hydrogen bonds in alcohols,
ethers, aldehydes, amines, hydrocarbons, etc.; silicon-
hydrogen, silicon-silicon, and silicon-carbon bonds; and
carbon-metal bonds of various types.

Carbon-carbon double bonds are usually the most reactive
of the above systems toward ozone, but nucleophiles such as
amines and selenides as well as certain carbon-nitrogen
double bonds are nearly as reactive or, in some cases, more
reactive. Carbon-hydrogen and silicon-hydrogen, etc., bonds
are usually the least reactive of the preceding groupings,
but do react, even below room temperature, provided a more
reactive grouping is not present in the substrate being
ozonized.

CARBON-CARBON DOUBLE BONDS: OZONOLYSIS

The classical reaction of ozone with organic molecules
is the ozonolysis reaction of carbon-carbon double bonds,
which was developed during the early part of the present
century by Harries.[9] Our present-day understanding of
this reaction is based on the extraordinarily detailed and
precise investigations of Criegee.[10,11] The Criegee mechanism of ozonolysis, refined to meet certain present day
requirements, is outlined in Figure 6.

The original Criegee mechanism did not specify the type
or size (3-, 4-, or 5-membered ring) of ozone-olefin adduct,
nor the mechanism by which it is produced. Today, the
adduct, or "initial ozonide," is thought in most cases to
be a 5-membered ring (III),[12,13,14] called a 1,2,3-trioxolane
and produced by a 1,3-dipolar cycloaddition.[3,15] The 5-
membered ring has been established in several instances by
nmr spectral studies.[12,13] In certain cases, however, it
is possible that the adduct is a 3-membered ring or a "σ
complex."[16,17]

The key intermediate in the Criegee mechanism is the
carbonyl oxide zwitterion (IV) produced by a reverse
1,3-dipolar cycloaddition of the initial ozonide (III).
Criegee originally wrote the zwitterion simply as $R_2\overset{+}{C}\text{-O-}\overset{-}{O}$,
but it now appears that it has considerable double bond

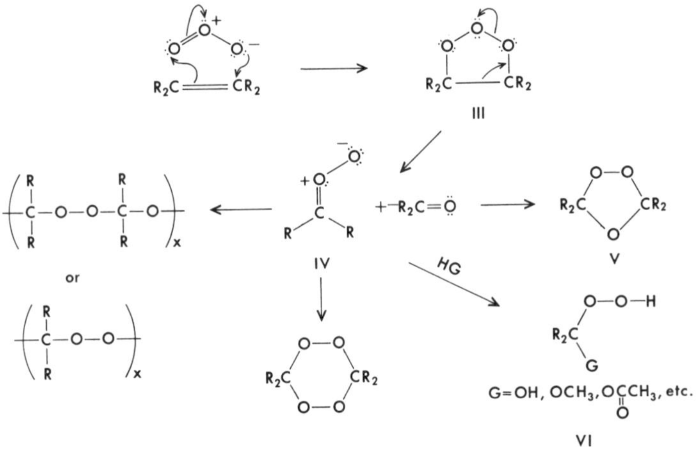

Figure 6. Criegee Ozonolysis Mechanism (Revised)

character and can exist in the form of <u>syn</u> and <u>anti</u> isomers.[15] The fate of the Criegee zwitterion depends on the structure of the original olefin and the reaction conditions employed. The possibilities are outlined in Figure 6; detailed discussions of the factors which favor each can be found elsewhere.[11] Of most importance to the present discussion is the reaction of the zwitterion (IV) with a protic solvent to give a hydroperoxide (VI). In water media, a hydroxy hydroperoxide (VI, G=OH) is the expected peroxidic ozonolysis product.

During the last several years, some interesting and novel studies, dealing especially with the stereochemistry of ozonide (V) formation, have cast doubt upon the Criegee mechanism as the sole mechanism of ozonolysis.[16] In one study, the highly unlikely suggestion was made that a 4-membered cyclic olefin-ozone adduct is a precursor to the 5-membered ring (or can go directly to ozonides).[18] Significantly, all of the new mechanisms acknowledge that the Criegee mechanism plays an essential, although perhaps sometimes a minor (in the percentage sense), role in ozonolysis, either as the key first step or as a competing reaction. From the viewpoint of this writer, it seems clear that there are competing mechanistic pathways in the ozonolysis reaction, but under the usual ozonolysis conditions the Criegee zwitterion mechanism, with recent stereochemical refinements,[15] is the major route to peroxidic ozonolysis products. Others may become predominant

under specialized conditions, but they are minor or non-
existant otherwise.

Ozonolyses in water media have been given very little
study. Pryde[19] has ozonized methyl oleate and methyl
lineoleate in water media and decomposed the peroxidic
ozonolysis products to aldehydes and/or carboxylic acids
under various conditions. His results are best explained
on the basis of a hydroxy hydroperoxide intermediate (VII,
Figure 7) which can either split out hydrogen peroxide to
give an aldehyde (VII → VIII) or dehydrate to a carboxylic
acid (IX → X). Criegee[20] has characterized such a
peroxide (XII) from ozonolysis of cyclic sulfone (XI) in
the presence of water. Fields[21,22] and co-workers have
ozonized various olefins and cycloolefins in aqueous
emulsions. However, either hydrogen peroxide or hydrogen
cyanide was also present, and the results are more interesting in regard to synthesis than to water treatment.

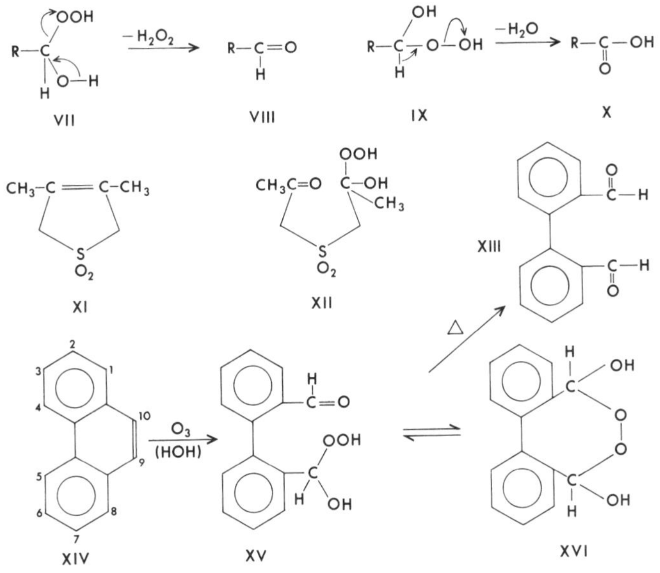

Figure 7. Ozonolyses in Water Media

Sturrock[23] has ozonized the aliphatic 9,10 bond of phenanthrene (XIV) in an aqueous t-butyl alcohol medium and obtained dialdehyde XIII upon steam distillation. By analogy to the ozonolysis of phenanthrene in methanol,[24] Sturrock suggested that the peroxidic ozonolysis product is predominantly XVI which is in equilibrium with XV. Loss of hydrogen peroxide from XV would give XIII. The results of both Pryde[19] and Sturrock,[23] however, indicate that decomposition of the peroxidic ozonolysis product (VII or equivalent) does not occur appreciably without heating. Further ozonation of the peroxide function of VII, XII and XV (or XVI) might occur slowly to give the corresponding aldehyde or ketone (via the hydrate).[25] Ozone should slowly convert the aldehydes to carboxylic acids, but additional ozone attack would probably be too slow to consider.

The literature gives numerous examples of "abnormal" ozonolyses in which the products contain fewer than the expected number of carbon atoms.[11] At present, it appears that all of these reactions involve either overozonation or conditions which favor a Baeyer-Villiger type rearrangement of peroxidic ozonolysis products.[26,27]

OLEFINIC DOUBLE BONDS: EPOXIDE FORMATION

With certain hindered olefins, another reaction competes with ozonolysis, in which the double bond is only partially cleaved and the product is an epoxide or rearrangement product thereof.[28] The reaction is illustrated in Figure 8 with the ozonation of 1-mesityl-1-phenylethylene (XVII), from which can be isolated either the epoxide (XX) or a mixture of the corresponding vinyl alcohol (XXI) and aldehyde (XIX). The latter compounds are produced by rearrangement of the epoxide (XX) via pathways 1) and 2).[28] Further ozonation is, of course, possible, especially with the vinyl alcohol (XXI).

The literature contains numerous examples of such epoxidations by ozone,[11,28] but no systematic study of the competition between this reaction and ozonolysis was made until 1967. It was then shown that as the bulk of the substituents at one carbon atom of the double bond of an olefin increased, the amount of epoxidation increased at the expense of ozonolysis.[28] A third bulky substituent also caused an increase in the extent to which epoxidation occurred in place of ozonolysis.[29] These results are illustrated in Table 4 which records the percentage of

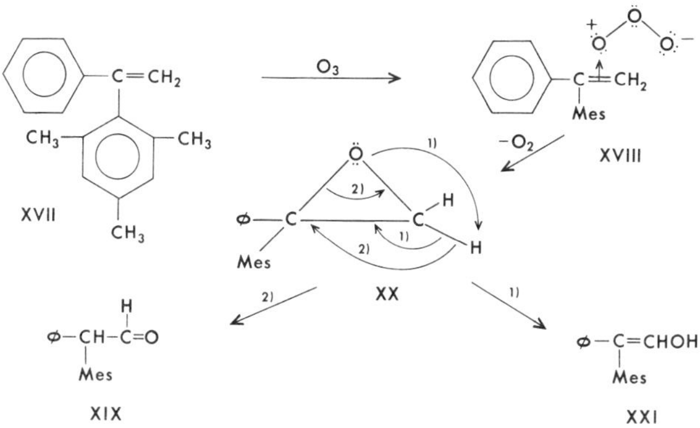

Figure 8. Epoxidation by Ozone

epoxidation occurring, the remainder being ozonolysis.
With aliphatic substituents, the product was usually the
epoxide,[28,29] while with aromatic substituents rearrangement of the epoxide tended to occur as illustrated in
Figure 8.[28]

As an explanation for this competition, it has been
suggested[28-30] that a π complex (XVIII) is the first
species produced between ozone and an olefin (or aromatic
molecule). In ozonolysis, this proceeds into the 1,3-
dipolar cycloaddition reaction to give the initial ozonide
(III, Figure 6) and ozonolysis products. As the bulk of
the substituents around a double bond increases, however,
the rate of 1,3-dipolar cycloaddition decreases and the π
complex, alternatively, decomposes to an epoxide, by loss
of molecular oxygen. Recently, π complexes have actually
been observed and characterized at very low (-150° to
-195° C) temperatures.[29,30] The epoxidation reaction
has not been studied in aqueous media. However, there
does not appear to be an appreciable solvent effect in
the competition between epoxidation and ozonolysis,[28]
which would indicate that the principles discussed would
also apply to aqueous ozonations.

Table 4. Competition between Ozonolysis and Epoxidation
        during Ozonation of Hindered Olefins

Olefin	Product	Yield	Reference
t-BuC = CH$_2$ | CH$_3$	epoxide[a]	10%	28
(t-Bu)$_2$C = CH$_2$	epoxide[a]	15%	28
(neopentyl)$_2$C = CH$_2$	epoxide	30%	29
(neopentyl)$_2$C = CH-t-Bu	epoxide	90%	29
$\emptyset_2$C = CH$_2$[d]	epoxide[a]	15%	28
$\emptyset$-C = CH$_2$[d] CH$_3$	epoxide[a]	30%	28
$\emptyset$C = CH$_2$[d] C -OH O	O = C$\overset{O}{-}$C$\overset{CH_2OH}{\underset{\emptyset}{-}}$[b,d]	40%	28
$\emptyset$C = CH$_2$[d] CH$_3$—⬡—CH$_3$ CH$_3$	XX, or XXI + XIX (Figure 8)	85%	28
Mes$_2$C = CH$_2$[c]	epoxide or vinyl alcohol	97%	29

[a]Only the molecular oxygen accompanying epoxide
formation was actually measured.

[b]Formed by interaction of the carboxyl group with the
epoxide ring.

[c]Mes = mesityl = $-C_6H_2(CH_3)_3$-2,4,6

[d]ϕ = phenyl = $-C_6H_5$

ACETYLENIC TRIPLE BONDS

Acetylenic compounds are known to react readily with ozone at the triple bond,[9],[11] but very few mechanistic or detailed studies of the reaction have been made.[31],[32] An application of the Criegee mechanism to triple bond ozonolysis[11] is shown in Figure 9. The structure of the proposed initial adduct (XXII) is unknown, but it presumably breaks down to a carbonyl oxide zwitterion (XXIII), from which peroxidic and nonperoxidic ozonolysis products can be derived, as shown in Figure 9. Ozonolyses of acetylenic compounds in aqueous media have not been studied, but, presumably, peroxide XXIV (R'=H) would be produced and would readily decompose to carboxylic acids as shown.

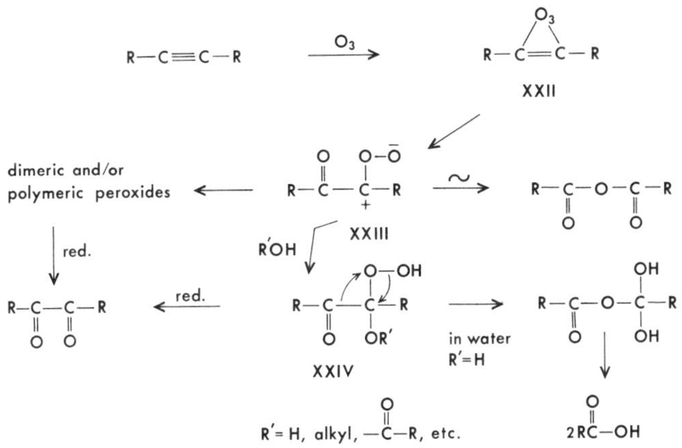

Figure 9. Ozonolysis of Acetylenic Compounds

AROMATIC COMPOUNDS: GENERAL

Ozonation of aromatic compounds appears to involve both 1,3-dipolar cycloaddition at a carbon-carbon bond, to give ozonolysis products, and electrophilic ozone attack at individual carbon atoms.[11] In regard to ease of ozone attack, the unsubstituted benzene ring is much less reactive than is an olefinic double bond.[11] Polycyclic aromatics such as phenanthrene, anthracene, and naphthalene

fall in between in reactivity.[11] Alkyl and other substituents which activate for electrophilic attack facilitate
ozone attack while those which deactivate for electrophilic
attack drastically decrease the rate of ozone attack on the
aromatic nucleus.[11] Peroxidic ozonolysis products have not
been characterized from ozonations of benzene or derivatives
thereof, due primarily to their instability. Benzene[33]
and its homologs,[11] however, appear to react with three
mole-equivalents of ozone, with the first mole attacking
any one of the six carbon-carbon bonds, to give the expected
ozonolysis products (glyoxals etc.). Very little investigation of these ozonolyses in aqueous media has been
reported.[33]

PHENOLS

Considerable study has been given recently to the
removal of phenolic wastes from water.[34] Phenols are more
reactive toward ozone than are most aromatics, and phenol
itself has been oxidized to carbon dioxide, formic acid,
glyoxal, and oxalic acid.[35] Eisenhauer[34],[36] has carried
out a detailed study of the ozonation of phenol (XXV,
Figure 10) in water solution and has identified catechol
(XXVIII) and o-quinone (XXXIV) as intermediary products,
although he did not establish that the reaction course
went solely through these intermediates. Muconic acid
(XXXV) was assumed to be the next intermediate, followed
by further ozonolysis of this unsaturated substance.[34]
From this author's viewpoint, it is quite possible that
a major course of the ozonation of phenol (XXV, Figure 10)
involved oxidation to catechol (XXVIII), but via electrophilic ozone attack (XXVI) rather than the cyclic intermediate proposed by Eisenhauer.[36] 1,3-Dipolar cycloaddition
(XXVII) should give rise directly to muconic acid (XXXV),
or the corresponding aldehydic acid (XXXIII), through
peroxidic intermediate XXX. This most likely does occur,
but to a minor extent. The suggested oxidation of catechol
(XXXVIII) to o-quinone (XXXIV, via XXXI) probably is only
a minor reaction, since it has been shown that catechol
undergoes a high conversion to muconic acid during ozonation.[37] Oxidation of o-quinone (XXXIV) to muconic acid
by ozone would require a nucleophilic ozone attack (XXXVI)
and a Baeyer-Villiger type rearrangement (XXXVI → XXXVII),
which probably would occur more slowly than attack on the
double bond system of the quinone (XXXIV). The more likely

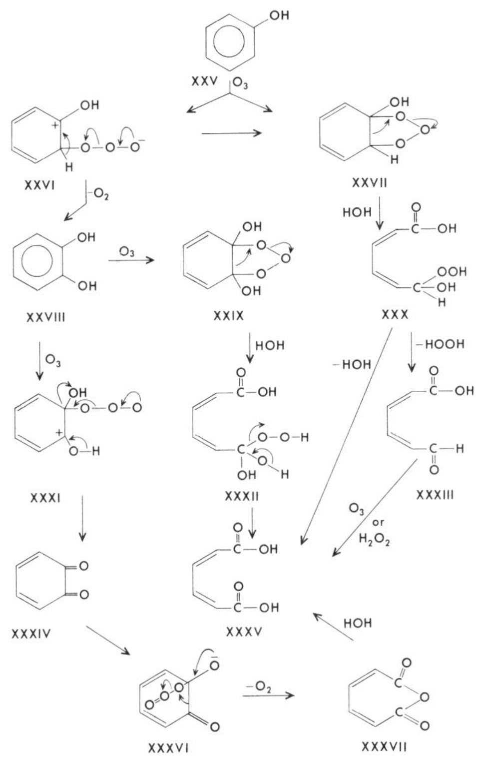

Figure 10. Ozonation of Phenol

reaction course for catechol ozonation would appear to be
XXVIII → XXIX → XXXII → XXXV.

Ozonation of other phenols and naphthols also have been
shown to occur readily.[37-39]

POLYCYCLIC AROMATICS

The ozonation of polycyclic aromatic hydrocarbons has
been studied extensively,[11,40,41] although most of the
work has been done in nonaqueous media. Three types of
ozone attack can occur, depending on the system: (1) 1,3-
dipolar cycloaddition (ozonolysis) at the bond or bonds

having the most double bond character (the bond with the lowest bond-localization energy); (2) electrophilic ozone attack where there are atoms of low atom-localization energy; (3) conjugate addition where there is a reactive diene system (a system of low para-localization energy).[40]

The most thorough studies have been made with naphthalene,[42,43] phenanthrene,[23,24] anthracene,[44] and benz[a]-anthracene.[40]   The ozonation of phenanthrene has already been discussed since ozonolysis at the 9,10-bond was the sole reaction (Figure 7); this bond has a very low bond-localization energy,[40] and is often described as having "four-fifths double bond character."

It is of especial interest to ozonation in aqueous media that Dobinson[45] has reported that ozonation of sodium 9-phenanthrenecarboxylate (XXXVIII, Figure 11) in water solution gave a 67% yield of phenanthrenequinone (XXXIX). The logical mechanism of this novel ozonation is outlined in Figure 11.

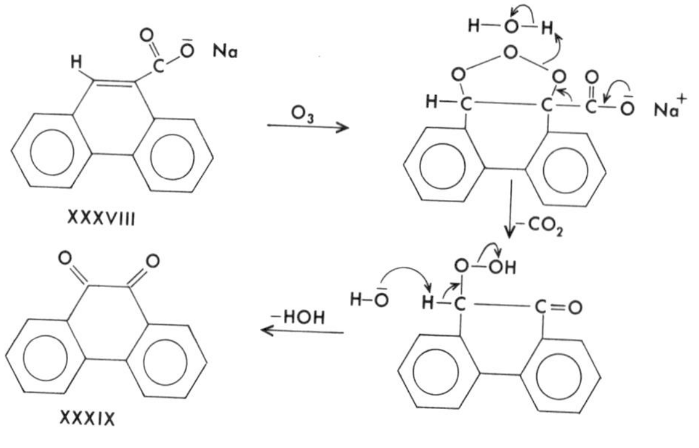

Figure 11. Ozonation of Sodium 9-Phenanthrenecarboxylate

Naphthalene (XL) also appears to undergo only ozonolysis, at the reactive 1,2 and 3,4 bonds.[42,43]   In methanolic solution, the cyclic product XLIIIb is formed (Figure 12), which can be decomposed to phthalaldehydic acid (XLV).[42] In aqueous acetone or aqueous t-butyl alcohol, the reaction apparently stops at stage XLI from which phthalaldehyde (XLII) can be derived.[43]   Further ozonation probably can occur to give phthalic acid (XLIV), but ozonation past this stage should be very slow.

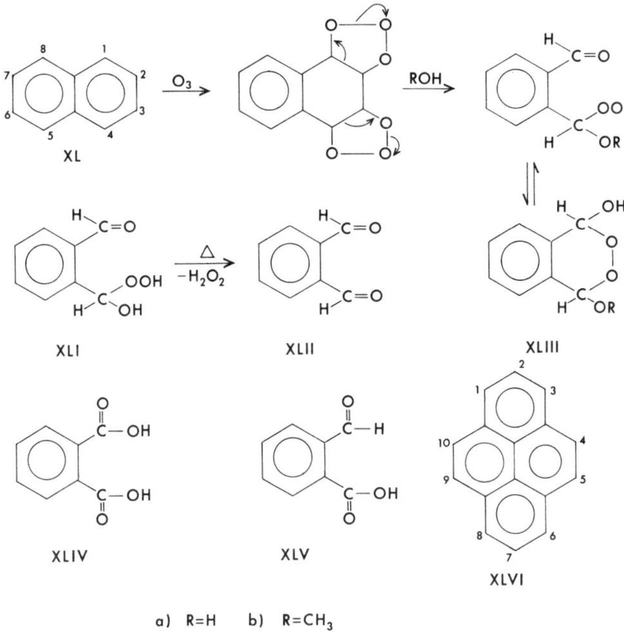

a) R=H   b) R=CH₃

Figure 12. Ozonation of Naphthalene

Pyrene (XLVI) has been ozonized in various media,[11] including aqueous t-butyl alcohol.[46] Ozonolysis occurs first at the reactive 4,5-bond, but the final products are complex.

Ozonation of anthracene and various benzanthracene types occurs via all three pathways mentioned at the start of this section.[11,40,41,44] Details will be given only for the anthracene ozonation.[44] Due to its insolubility, anthracene could not be ozonized in water alone. However, using aqueous acetic acid solvent, a profound solvent effect was found, at least in comparison to aprotic solvents.[44] Ozonation of anthracene (XLVII, Figure 13) in chlorinated hydrocarbon solvents gave up to a 67% yield of phthalic acid (XLIV) as the major product, after an oxidative work-up procedure, and only 25 to 35% anthraquinone (LV) yields. In 90% acetic acid, the results were just the opposite, with anthraquinone (LV) yields as high as 73%.[44] Ozonation in aqueous acetic acid resulted in

the precipitation of 35 to 40% yields of anthraquinone
(LV) during the ozonation; this was accompanied by the
evolution of three moles of molecular oxygen per mole of
anthraquinone. The remaining 30 to 35% anthraquinone
yield was obtained by dehydration of a peroxidic intermediate.[44] These results can be explained by the mechanism
outlined in Figure 13.

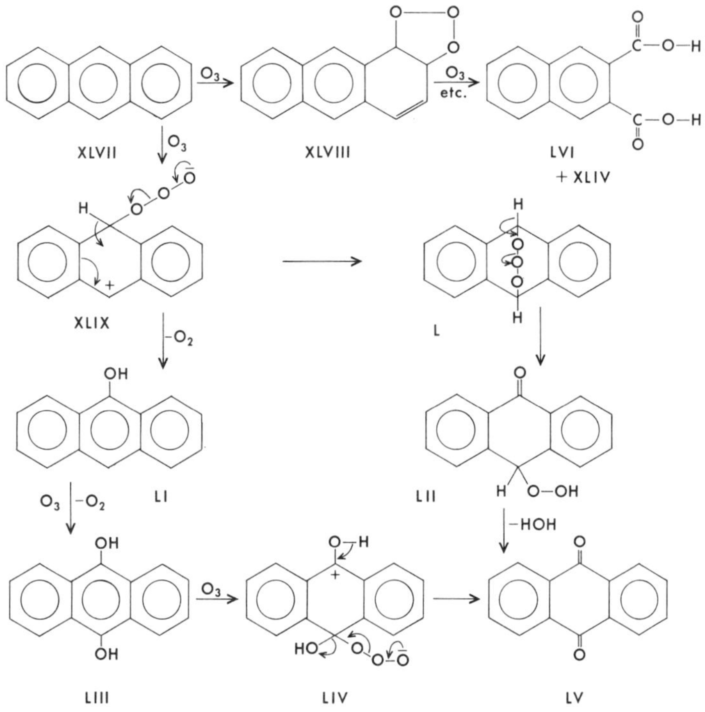

Figure 13. Ozonation of Anthracene

Roughly 25% of the ozonation involves 1,3-dipolar cycloaddition, beginning with XLVIII and continuing at the outer
ring and on into the center ring to give, after oxidative
workup, naphthalenedicarboxylic acid (LVI) and/or phthalic
acid (XLIV). The approximately 40% anthraquinone yield
which precipitates during the ozonation arises from three
consecutive electrophilic ozone attacks accompanied by

molecular oxygen evolution (XLVII → XLIX → LI → LIII → LIV → LV). Intermediate XLIX can also complete a conjugate addition to give transannular ozonide L.[44] Whereas this intermediate is highly unstable in the case of anthracene, the analogous adduct was actually isolated and characterized in the case of 9,10-dimethylanthracene.[47] It is with intermediate L that the major solvent effect is thought to occur.[44] In a protic, hydroxylic solvent such as aqueous acetic acid, a smooth rearrangement of L to LII occurs, aided by the donation and acceptance of protons by the solvent. Dehydration of LII gives anthraquinone (LV). In an aprotic, nonhydroxylic solvent like carbon tetrachloride, L rearranges via ring cleavage to give intermediates which can be oxidized to phthalic acid (XLIV). Anthraquinone is resistant to further ozonation, except very slowly.

Sturrock and co-workers[48] have reported in the patent literature that ozonation of phenanthrene-anthracene mixtures in 30% t-butyl alcohol-70% water results in yields of anthraquinone as high as 72%.

Similar competitions occur during ozonation of benz[a]anthracene[40] and other anthracene-type polycyclic aromatics.[40,41] The relative importance of ring cleavage (ozonolysis) versus quinone formation depends on the relative values of the bond-localization energies versus atom- and/or para-localization energies.[40]

HETEROCYCLICS

Most aromatic and aliphatic unsaturated heterocycles are readily attacked by ozone.[11] An exception is the pyridine ring which reacts extremely slowly.[11] Quinolines are preferentially attacked in the carbocyclic ring.[11] Pyrroles[11] and furans[11,49] are quite reactive toward ozone, and indoles and benzofurans are readily attacked at the heterocyclic carbon-carbon "double bond."[11]

An interesting solvent effect has been found with 2,5-diarylfurans (LVII, Figure 14) involving water and other protic, hydroxylic solvents.[49,50] Two competing types of ozone attack are involved, similar to those discussed with anthracene types and hindered olefins. The major reaction, as outlined in Figure 14, is 1,3-dipolar cycloaddition (e.g., LIX) at the 2,3- and/or 4,5-bonds to give ozonolysis products such as benzoic anhydride, benzoic acid, formic acid, etc. (when Ar in Figure 14 is phenyl).[49] Competing

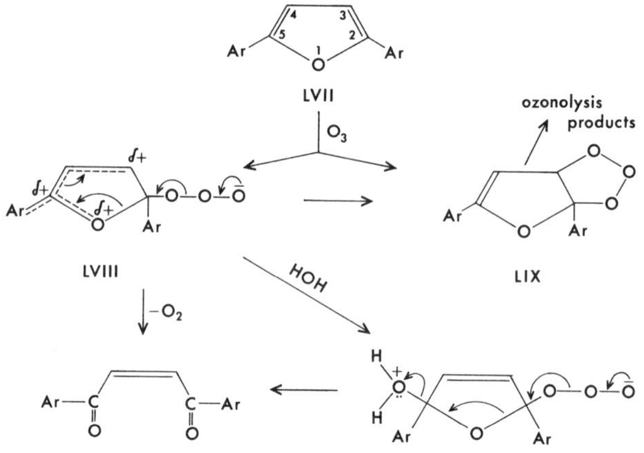

Figure 14. Ozonation of 2,5-Diarylfurans

with this is electrophilic ozone attack (LVIII) at the
reactive 2 or 5 positions. This can either lose oxygen
and yield cis-1,2-diaroylethylenes (LX) or complete a
cycloaddition (to LIX). Ozonations of 2,5-diphenylfuran
(LVII, Ar = phenyl) at room temperature yielded only 10
to 11% cis-1,2-dibenzoylethylene (LX) in methylene chloride
solvent, whereas in 90% aqueous acetic acid the yield of
LX was 33%. The probable explanation is that in the absence of a hydroxylic solvent, LVIII tends to cyclize to
LIX in preference to losing molecular oxygen, while in
the presence of water (or other hydroxylic solvents) LVIII
is trapped in the form of LXI, which then loses oxygen and
water to yield LX.[49] Further ozonolysis of cis-1,2-
diaroylethylenes (LX) affords the corresponding arylglyoxal
and/or oxidative decomposition products thereof.[49]

NUCLEOPHILES

Various organic substances with a nucleophilic atom in
their structures are readily oxidized by ozone.[11] Examples
of such oxidations include the conversion of sulfides to
sulfoxides; sulfoxides to sulfones; tertiary amines, phosphines, and arsines to the corresponding oxides;[11]
phosphites to phosphates;[51] selenides to selenoxides;[52]

di- and polysulfides to sulfonic anhydrides;[11] diselenides
to selenic anhydrides;[52] and primary and secondary amines
to nitro compounds, nitroxides, and various decomposition
products.[53-56] Some of these are illustrated in Figure 15.

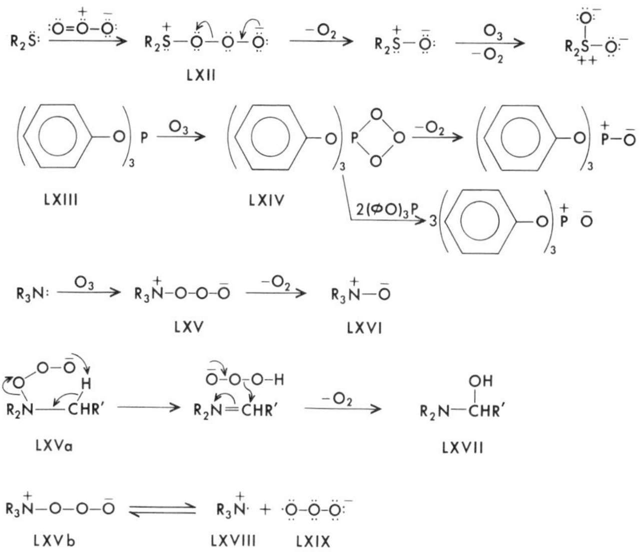

Figure 15. Ozonation of Nucleophiles

In some cases, especially with certain amines[56] and
selenides,[52] the ozonation is faster than that of olefinic
double bonds. In all of these cases, the initial ozone
attack is thought to be electrophilic, producing an ozone
adduct, as illustrated in Figure 15 (LXII, LXIV, LXV).
The adduct proved stable enough to be characterized in the
case of triphenyl phosphite (LXIII → LXIV),[51] but this has
not yet been accomplished in the other cases.

The most detailed study in the ozonation of nucleophiles
area has been with primary, secondary, and tertiary
amines,[53-55,57] which has given evidence for at least four

fates of the amine-ozone adduct (LXV). These are
illustrated in Figures 15 and 16.

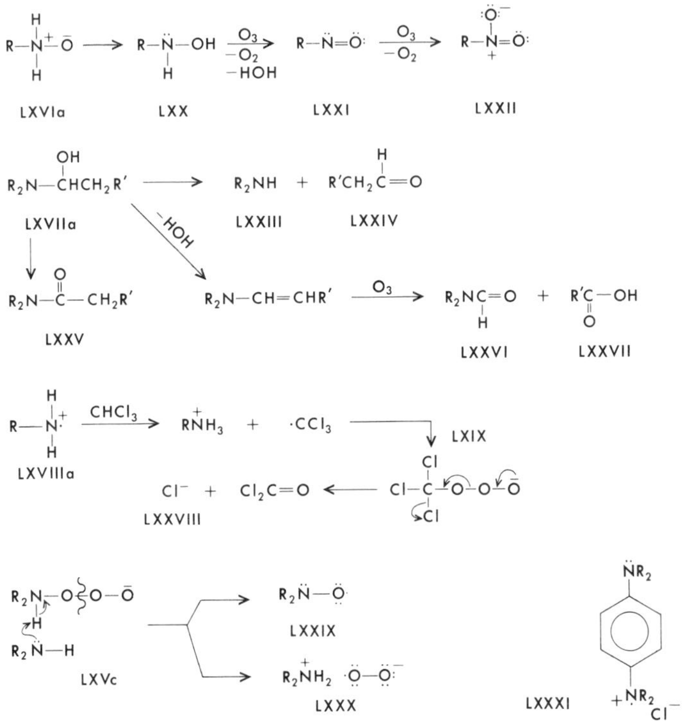

Figure 16. Ozonation of Amines

Loss of oxygen from the adduct yields an amine oxide,
which is isolable in the case of tertiary amines[57] (LXV →
LXVI), but reacts further in the case of primary and
secondary amines.[54,55] With primary amines the amine
oxide rearranges to a hydroxylamine (LXX) which is similarly oxidized further, first to the nitroso (LXXI) and
then to the nitro (LXXII) compounds.[54] Competing with
amine oxide formation, especially with amines possessing
primary alkyl groups, is side-chain oxidation which is
thought to occur predominantly by the route illustrated

in Figure 15 (LXVa → LXVII) followed by further reactions, including oxidation, of amino alcohol LXVII. This is illustrated in Figure 16 (LXVIIa → LXXIII, LXXIV, LXXV, LXXVI, LXXVII). It is obvious that deep-seated oxidation can result from side-chain ozonation, especially when dissociation of LXVIIa to a lower amine and an aldehyde occurs, followed by further ozonation of these species. A solvent effect influencing the competition between amine oxide formation and side-chain ozonation has been observed with certain tertiary amines.[57,58] However, in aqueous medium it appears that side-chain oxidation is the major reaction.[58]

A third fate of the amine-ozone adduct (LXVb) is dissociation to an amine cation radical (LXVIII) and the ozonate anion radical (LXIX). This reaction tends to be a major route, however, only with primary amines in chlorinated solvents, in which case ammonium chlorides are produced as illustrated in Figure 16 (LXVIIIa → LXXVIII).[54]

The fourth fate is predominant with secondary amines.[55] It involves interaction of the amine-ozone adduct (LXVc, Figure 16) with a second molecule of the amine to give a dialkyl nitroxide (LXXIX) and an ammonium superoxide (LXXX). The latter goes to chloride in chlorinated hydrocarbon solvents. The nitroxide (LXXIX) undergoes further ozonation to give the corresponding nitroalkane and oxidation products of the expelled alkyl group.[59]

Further investigation of the ozonation of amines in aqueous media is needed. Besides the ozonation of tri-n-butylamine in water suspension,[58] the only ozonation of a simple amine in water reported appears to be the early study of Strecker and Baltes[60] with trimethylamine. In both cases, side-chain attack was the major reaction. It is possible, however, that this occurred by 1,3-dipolar insertion (to be discussed, see Figure 18) rather than by the mechanism just described, since electrophilic ozone attack to give an amine-ozone adduct may be slowed down by hydration of the amine. In acidic solutions, amines are unreactive toward ozone[61] due to the unavailability of the nitrogen electron pair, which would not only prevent electrophilic ozone attack but also slow down 1,3-dipolar insertion.

Various amino acids and proteins have been ozonized in water solution, but the ozone attack appears to have occurred at sulfur (cystine) or aromatic or heterocyclic unsaturated carbon-carbon bonds rather than nitrogen,[62] due to the inner salt structures involved. Tetra-N-substituted phenylenediamines react extremely fast with

ozone even in acidic aqueous solutions to give Würster salts
(LXXXI).[63] The driving force here seems to be the formation
of the resonance-stabilized cation radical.

Ammonia is also easily oxidized by ozone to nitrates, but
the reaction is somewhat slow in water solution for reasons
already discussed.[64] Also, some of the ammonia ends up as
the ammonium cation.[64]

CARBON-NITROGEN DOUBLE BONDS, etc.

The carbon-nitrogen double bond of Schiff bases (imines),[5,65]
2,4-dinitrophenylhydrazones[66] and other hydrazones,[65]
oximes,[65] nitrones,[5,67] azines,[68] and substituted diazomethanes[69,70] readily reacts with ozone. The major reaction
course involves cleavage of the carbon-nitrogen double bond
to yield the corresponding ketone (or aldehyde and/or
carboxylic acid). With Schiff bases (LXXXII, Figure 17)
an oxaziridine (LXXXIII) and an amide (LXXXIV) are also
often produced.[5] These reactions can be explained by
either an electrophilic or a nucleophilic ozone attack.
The evidence appears to favor the electrophilic attack in
most cases.[65] However, from the viewpoint of the present
writer, the nucleophilic ozone attack best expalins the
results with Schiff bases.[5] It is also possible that
Schiff bases of aldehydes and monosubstituted diazomethanes
react, at least partially, by ozone insertion at the
carbon-hydrogen bond,[70] as will be described in the next
section for aldehydes.

Figure 17 illustrates a nucleophilic ozone attack on a
Schiff base (LXXXII) and an electrophilic ozone attack on
a dimethylphenylhydrazone (LXXXV). It is of interest that
the latter. compound ozonizes at approximately the same
rate as does trans-stilbenze (an olefin).[65]

Although the ozonation of alkyl isocyanides (LXXXVI)
to isocyanates (LXXXIX) has been reported,[71] no comparable
ozonation of nitriles (R-C≡N:) appears to have been
studied. However, inorganic cyanides in water solution
are rapidly converted to cyanates by ozone.[72] The iso-
cyanide ozonation could involve either an electrophilic
(LXXXVII) or a nucleophilic (LXXXVIII) ozone attack. The
cyanide anion ozonation probably involves electrophilic
attack similar to that with amines.

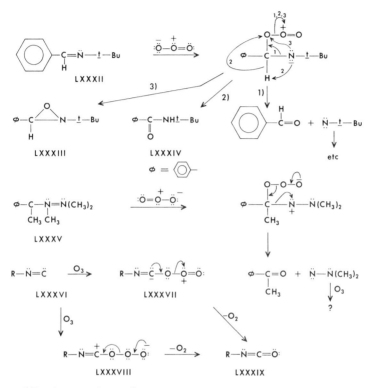

Figure 17. Ozonation of Carbon-Nitrogen Double Bonds

OZONATION OF CARBON-HYDROGEN BONDS

When reactive groupings of the types discussed in the preceding sections are not present in organic molecules, ozone attack on carbon-hydrogen bonds becomes possible. This reaction occurs readily with aldehydes,[73-75] ethers,[76,77] alcohols,[78] and even hydrocarbons[79,80] or hydrocarbon groupings[81] having secondary and tertiary carbon-hydrogen bonds. By these reactions, aldehydes are converted to carboxylic acids, primary and secondary alcohols to carboxylic acids and/or aldehydes or ketones, ethers to alcohols and esters, and hydrocarbons to alcohols and ketones.

There appears to be general agreement today that these ozonations involve a hydrotrioxide (e.g., XCI) intermediate. However, the exact mode of formation of this intermediate and how it is converted to products is not agreed upon. From the viewpoint of this reviewer, a 1,3-dipolar insertion of ozone, in which there is developing carbonium ion character in the transition state (XC), best explains the results, as illustrated in Figure 18. By such a mechanism

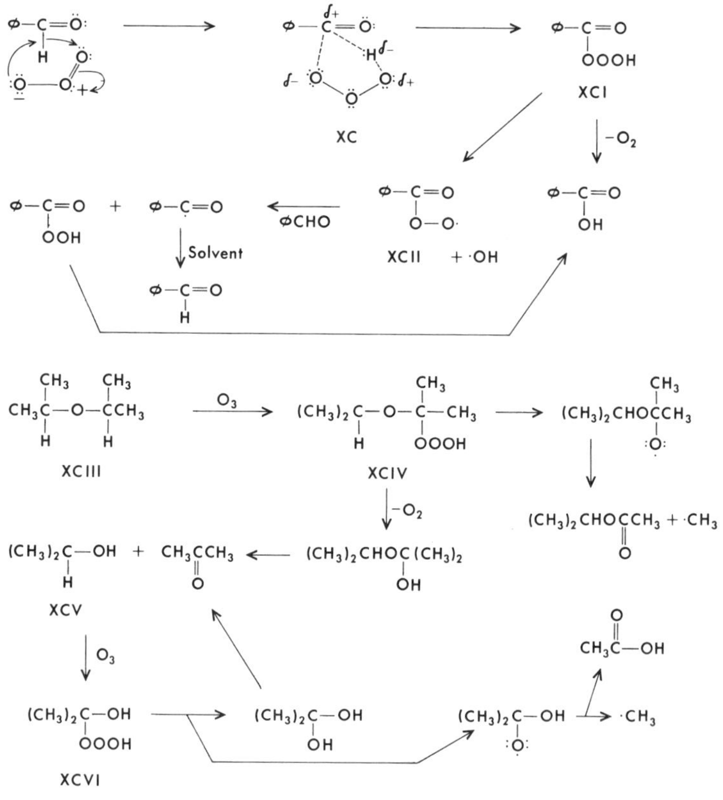

Figure 18. Carbon-Hydrogen Bond Ozonations

electron-withdrawing groups would be expected to slow down the ozonation, due to destabilization of the transition state, as found.[73,81] This also explains nicely the 80 to 90% retention of configuration observed during ozonation of the cis- and trans-1,2-dimethylcyclohexanes to the corresponding alcohols,[80] as illustrated in Figure 19.

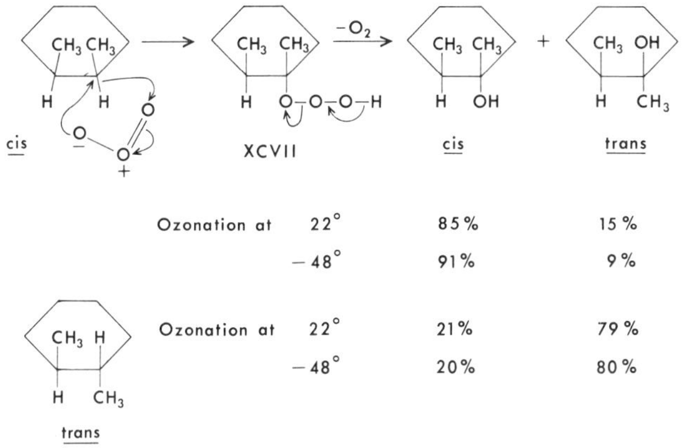

Figure 19. Stereochemistry of C-H Bond Ozonation

The hydrotrioxide intermediate (XCI, XCIV, XCVI, XCVII) can break down either by a radical pathway, as best illustrated in Figure 18 for the benzaldehyde ozonation (XCI → XCII etc.), or in a concerted manner, as shown in Figure 19, to yield singlet oxygen[82] and the corresponding hydroxyl compound. Some radical pathway must also occur with the dimethylcyclohexanes (Figure 19) since retention of configuration was not 100%. These reactions are also illustrated with ethers (XCIII) and alcohols (XCV) in Figure 18.

The ozonation of ethers has recently been carried over to acetals, including sugar glycosides, to yield esters.[83] The literature also records the ozonation of sugar alcohols and of polysaccharides in aqueous medium,[84] including the degradation of wood pulp.[85] Humic acids have been degraded by ozone in aqueous solution,[86] but nothing is known of the chemistry involved.

A simple carbon-hydrogen bond ozonation which has been performed in water solution is that of malonic acid.[87] The methylene group (CH_2) was converted to an alcohol and to a ketone function. Also, oxalic acid and carbon dioxide was produced. These reactions are understandable on the bases of the mechanisms already illustrated.

SILICON-HYDROGEN AND CARBON-METAL BONDS

Silicon-hydrogen bonds can be cleaved by ozone in a similar fashion, as just described for carbon-hydrogen bonds, to give the corresponding hydroxyl compounds.[88,89] Similarly, silicon-silicon and silicon-carbon bonds can be cleaved.[88]

Ozonation of alkylmercuric halides and dialkylmercurials gave good yields of carboxylic acids, ketones, and tertiary alcohols from primary, secondary, and tertiary alkylmer-curials, respectively.[90] Carbon-mercury bond cleavage was accompanied by varying degrees of carbon-carbon cleavage to give carboxylic acids of lower carbon content than of the alkyl groups involved. These reactions are thought to involve insertion at a carbon-mercury bond similar to those insertions already discussed.[90] The decomposition routes of the trioxy intermediate to the final products have not been elucidated. Alcohols are among the intermediate products and are oxidized further by routes already discussed. It seems likely that radical pathways are involved, resulting in carbon-carbon bond cleavage products.

SUMMARY

The intent of this brief résumé of the reactions of ozone with organic compounds and of the mechanisms of these reactions has been to call attention to the great versatility and power of ozone as an oxidizing agent. In the mind of this author, it is the great hope of the future in sewage and waste treatment and water purification.

Godspeed to those whose responsibility it is to adapt the above described reactions to this all-important and urgent practical use!

REFERENCES

1. Trambarulo, R., Ghosh, S. N., Burrus, C. A. Jr., and Gordy, W., "The Molecular Structure, Dipole Moment, and g Factor of Ozone from Its Microwave Spectrum." The Journal of Chemical Physics, *21*, 851 (1953).
2. Orville-Thomas, W. J., "A Bond-Order/Bond-Length Relation for Oxygen-Oxygen Bonds." Journal of Molecular Spectroscopy, *3*, 588 (1959).
3. Huisgen, R., "1,3-Dipolar Cycloadditions." Angewandte Chemie International Edition in English, *2*, 565 (1963).
4. Huisgen, R., "Kinetics and Mechanism of 1,3-Dipolar Cycloadditions." Angewandte Chemie International Edition in English, *2*, 633 (1963).
5. Riebel, A. H., Erickson, R. E., Abshire, C. J., and Bailey, P. S., "Ozonation of Carbon-Nitrogen Double Bonds. I. Nucleophilic Attack of Ozone." Journal of the American Chemical Society, *82*, 1801 (1960).
6. Benson, S. W., "Comments on the Mechanism of Ozone Rate Action with Hydrocarbons and Alcohols." Advances in Chemistry Series, No. 77. American Chemical Society, Washington, D.C. (1968), p. 74.
7. Murray, R. W., "Ozonolysis of Organic Compounds," in Denney, D. B., Techniques and Methods of Organic and Organometallic Chemistry. Marcel Dekker, New York (1969), p. 1.
8. Charlot, G., Bezier, D., and Courtot, J., Selected Constants. Oxydo-Reduction Potentials. Vol. 8 of Tables of Constants and Numerical Data. Pergamon Press, New York (1968), pp. 12, 21.
9. Long, L. Jr., "The Ozonization Reaction." Chemical Reviews, *27*, 437 (1940).
10. Criegee, R., "The Course of Ozonization of Unsaturated Compounds." Record of Chemical Progress (Kresge-Hooker Science Library Associates), *18*, 111 (1957).
11. Bailey, P. S., "The Reactions of Ozone with Organic Compounds." Chemical Reviews, *58*, 925 (1958).
12. Bailey, P. S., Thompson, J. A., and Shoulders, B. A., "Structure of the Initial Ozone-Olefin Adduct." Journal of the American Chemical Society, *88*, 4098 (1966).
13. Durham, L. J., and Greenwood, F. L., "Ozonolysis. X. The Molozonide as an Intermediate in the Ozonolysis of

cis and trans Alkenes." Journal of Organic Chemistry,
33, 1629 (1968).
14. Criegee, R. "Neues aus der Chemie der Ozonide." Chimia,
22, 392 (1968).
15. Bauld, N. L., Thompson, J. A., Hudson, C. E., and
Bailey, P. S., "Stereospecificity in Ozonide and Cross-
Ozonide Formation." Journal of the American Chemical
Society, *90*, 1822 (1968).
16. Murray, R. W., "The Mechanism of Ozonolysis." Accounts
of Chemical Research, *1*, 313 (1968).
17. Reinhardt, H. G., Doorakian, G. A., and Freedman,
H. H., "Steric Effects in the Ozonolyses of 1,2,3,4-
Tetraphenylcyclobutenes." Journal of the American
Chemical Society, *90*, 5934 (1968).
18. Story, P. R., Alford, J. A., Ray, W. C., and Burgess,
J. R., "Mechanisms of Ozonolysis. A New and Unifying
Concept." Journal of the American Chemical Society,
93, 3044 (1971).
19. Pryde, E. H., Moore, D. J., and Cowan, J. C., "Hydro-
lytic, Reductive and Pyrolytic Decomposition of Selected
Ozonolysis Products. Water as an Ozonization Medium."
The Journal of the American Oil Chemist's Society, *45*,
888 (1968).
20. Criegee, R., and Lohaus, G., "Die Ozonide ungesattigter
cyclischer Sulfone." Justus Liebig's Annalen der Chemie,
583, 1 (1953).
21. Fremery, M. I., and Fields, E. K., "Emulsion Ozonization
of Cycloolefins in Aqueous Alkaline Hydrogen Peroxide."
Journal of Organic Chemistry, *28*, 2537 (1963).
22. Fields, E. K., "Cyanozonolysis of Olefins." Advances in
Chemistry Series, No. 51. American Chemical Society,
Washington, D.C. (1965), p. 99.
23. Sturrock, M. G., Cline, E. L., and Robinson, K. R.,
"The Ozonation of Phenanthrene with Water as Participating Solvent." Journal of Organic Chemistry, *28*,
2340 (1963).
24. Bailey, P. S., "The Ozonolysis of Phenanthrene in
Methanol." Journal of the American Chemical Society,
78, 3811 (1956).
25. Barnard, D., McSweeney, G. P., and Smith, J. F., "The
Reaction of Ozone with Organic Hydroperoxides."
Tetrahedron Letters, 1 (1960).
26. Story, P. R., and Burgess, J. R., "The Baeyer-Villiger
Reaction as a Source of 'Abnormal' Ozonolysis Products."
Tetrahedron Letters, 1287 (1968).
27. Kolsaker, P., and Bailey, P. S., "Ozonation of Compounds
of the Type ArCH=CHG; Ozonation in Methanol." Acta
Chemica Scandinavica, *21*, 537 (1967).

28. Bailey, P. S., and Lane, A. G., "Competition between Complete and Partial Cleavage during Ozonation of Olefins." Journal of the American Chemical Society, *89*, 4473 (1967).
29. Bailey, P. S., Ward, J. W., Hornish, R. E., and Potts, F. E. III, "Complexes and Radicals Produced during Ozonation of Olefins." Advances in Chemistry Series,No.112. American Chemical Society, Washington, D.C. (1972), p. 1.
30. Bailey, P. S., Ward, J. W., and Hornish, R. E., "Complexes of Ozone with Carbon π Systems." Journal of the American Chemical Society, *93*, 3552 (1971).
31. Criegee, R., and Lederer, M., "Die Ozonspaltung der Acetylenbindung." Justus Liebig's Annalen der Chemie, *583*, 29 (1953).
32. Bailey, P. S., Chang, Yun-Ger, and Kwie, W. W. L., "Ozonolysis of Unsymmetrical Acetylenes." Journal of Organic Chemistry, *27*, 1198 (1962).
33. Keaveney, W. P., Rush, R. V., and Pappas, J. J., "Glyoxal from Ozonolysis of Benzene." Industrial and Engineering Chemistry Product Research and Development, *8*, 89 (1969).
34. Eisenhauer, H. R., "The Ozonation of Phenolic Wastes." Journal of the Water Pollution Control Federation, *40*, 1887 (1968).
35. Bernatek, E., and Frengen, C., "Ozonolysis of Phenols. I. Ozonolysis of Phenol in Ethyl Acetate." Acta Chemica Scandinavica, *15*, 471 (1961).
36. Eisenhauer, H. R., "Increased Rate and Efficiency of Phenolic Waste Ozonation." Journal of the Water Pollution Control Federation, *43*, 201 (1971).
37. Wingard, L. B. Jr., and Finn, R. K., "Oxidation of Catechol to cis, cis-Muconic Acid with Ozone." Industrial and Engineering Chemistry Product Research and Development, *8*, 65 (1969).
38. Bernatek, E., Moskeland, J., and Valen, K., "Ozonolysis of Phenols. II. Catechol, Resorcinol and Quinol." Acta Chemica Scandinavica, *15*,1454 (1961).
39. Bernatek, E., and Frengen, C., "Ozonolysis of Phenols. III. 1- and 2-Naphthol." Acta Chemica Scandinavica, *15*, 2421 (1962).
40. Bailey, P. S., Batterbee, J. E., and Lane, A. G., "Ozonation of Benz[a]anthracene." Journal of the American Chemical Society, *90*, 1027 (1968).
41. Moriconi, E. J., and Salce, L., "Ozonation of Polycyclic Aromatics." Advances in Chemistry Series, No. 77. American Chemical Society, Washington, D.C. (1968), p. 65.

42. Bailey, P. S., Bath, S. S., Dobinson, F., Garcia-Sharp, F. J., and Johnson, C. D., "Ozonolysis of Naphthalenes. The Aromatic Products." Journal of Organic Chemistry, *29*, 697 (1964).

43. Sturrock, M. G., Cravy, B. J., and Wing, V. A. "Ozonation of Naphthalene with Water as Participating Solvent. Preparation of o-Phthaldehyde." Canadian Journal of Chemistry, *49*, 3047 (1971).

44. Bailey, P. S., Kolsaker, P., Sinha, B., Ashton, J. B. Dobinson, F., and Batterbee, J. E., "Competing Reactions in the Ozonation of Anthracene." Journal of Organic Chemistry, *29*, 1400 (1964).

45. Dobinson, F., "Aqueous Ozonisations: Phenanthrene-9-carboxylic Acid." Chemistry and Industry, 1122 (1959).

46. Sturrock, M. G., and Duncan, R. A., "The Ozonation of Pyrene. A Monomeric Monoozonide formed in Polar Solvents." Journal of Organic Chemistry, *33*, 2149 (1968).

47. Erickson, R. E., Bailey, P. S., and Davis, J. C. Jr., "Structure of the Monoozonide of 9,10-Dimethylanthracene." Tetrahedron, *18*, 388 (1962).

48. Sturrock, M. G., Cline, E. L. and Robinson, K. R., "Ozonolysis of Coal-Tar Products." U.S. Patent 2,898,350 (August 4, 1959).

49. White, H. M., Colomb, H. O. Jr., and Bailey, P. S., "Ozonation of 2,5-Diphenylfuran." Journal of Organic Chemistry, *30*, 481 (1965).

50. Bailey, P. S., White, H. M., and Colomb, H. O. Jr., "Ozonation of Diarylfurans." Journal of Organic Chemistry, *30*, 487 (1965).

51. Thompson, Q. E., "Ozone Oxidation of Nucleophilic Substances. I. Tertiary Phosphite Esters." Journal of the American Chemical Society, *83*, 845 (1961).

52. Ayrey, G., Barnard D., and Woodbridge, D. T., "The Oxidation of Organoselenium Compounds by Ozone." Journal of the Chemical Society, 2089 (1962).

53. Bailey, P. S., Keller, J. E., Mitchard, D. A., and White, H. M., "Ozonation of Amines." Advances in Chemistry Series, No. 77. American Chemical Society, Washington, D.C. (1968), p. 58.

54. Bailey, P. S., and Keller, J. E., "Ozonation of Amines. III. t-Butylamine." Journal of Organic Chemistry, *33*, 2680 (1968).

55. Bailey, P. S., Keller, J. E., and Carter, T. P. Jr., "Ozonation of Amines. IV. Di-t-butylamine." Journal of Organic Chemistry, *35*, 2777 (1970).

56. Razumovskii, S. D., Buchachenko, A. L., Shapiro, A. B., Rozantsev, E. G., and Zaikov, G. E., "The Formation of Nitroxyl Radicals in the Reaction of Amines with Ozone." Doklady Chemistry, *183*, 1086 (1968).

57. Bailey, P. S., Mitchard, D. A., and Khashab, Abdul-Ilah Y., "Ozonation of Amines. II. The Competition between Amine Oxide Formation and Side-Chain Oxidation." Journal of Organic Chemistry, *33*, 2675 (1968).

58. Henbest, H. B., and Stratford, M. J. W., "Amine Oxidation. Part VIII. The Reaction of Tri-n-butylamine with Ozone." Journal of the Chemical Society, 711 (1964).

59. Bailey, P. S., and Keller, J. E., "Ozonation of Amines. V. Di-t-butyl Nitroxide." Journal of Organic Chemistry, *35*, 2782 (1970).

60. Strecker, W., and Baltes, M., "Über die Einwirkung von Ozon auf Aliphatische, und aromatische Substitutionsprodukte des Ammoniaks." Berichte der Deutschen Chemischen Gesellschaft, *54*, 2693 (1921).

61. Horner, L., Schaefer, H., and Ludwig, W., "Ozonisierung von tertiaren Aminen (Phosphinen und Arsinen) sowie Thioathern und Disulfiden." Chemische Berichte, *91*, 75 (1958).

62. Mudd, J. B., Leavitt, R., Ongun, A., and McManus, T. T., "Reaction of Ozone with Amino Acids and Proteins." Atmospheric Environment, *3*, 669 (1969).

63. Layer, R. W., "Reaction of Ozone with p-Phenylenediamine and Related Compounds." Rubber Chemistry and Technology, *39*, 1584 (1966).

64. Papko, S. I., "Action of Certain Heterogeneous Catalysts on the Oxidation of Ammonia in Aqueous Solution by Ozonized Oxygen." Journal of Applied Chemistry of the USSR, *30*, 1361 (1957).

65. Erickson, R. E., Andrulis, P. J. Jr., Collins, J. C., Lungle, M. L., and Mercer, G. D., "Mechanism of Ozonation Reactions. IV. Carbon-Nitrogen Double Bonds." Journal of Organic Chemistry, *34*, 2961 (1969).

66. Erickson, R. E., Riebel, A. H., Reader, A. M. and Bailey, P. S., "Ozonisierung von 2,4-Dinitrophenylhydrazonen." Justus Leibig's Annalen der Chemie, *653*, 129 (1962).

67. Erickson, R. E., and Myszkiewicz, T. M., "Mechanism of Ozonation Reactions. Nitrones." Journal of Organic Chemistry, *30*, 4326 (1965).

68. Miller, R. E., "Ozonation of Azo and Azomethine Double Bonds." Journal of Organic Chemistry, *26*, 2327 (1961).

69. Reader, A. M., Bailey, P. S., and White, H. M., "Ozonation of Substituted Diazomethanes." Journal of Organic Chemistry, *30*, 784 (1965).

70. Bailey, P. S., Reader, A. M., Kolsaker, P., White, H. M., and Barborak, J. C., "Ozonation of Monosubstituted Diazomethanes." Journal of Organic Chemistry, *30*, 3042 (1965).
71. Feuer, H., Rubinstein, H., and Nielsen, A. T., "Reaction of Alkyl Isocyanides with Ozone. A New Isocyanate Synthesis." Journal of Organic Chemistry, *23*, 1107 (1958).
72. Selm, R. P., "Ozone Oxidation of Aqueous Cyanide Waste Solutions in Stirred Batch Reactors and Packed Towers." Advances in Chemistry Series, No. 21. American Chemical Society, Washington, D.C. (1959), p. 66.
73. White, H. M., and Bailey, P. S., "Ozonation of Aromatic Aldehydes." Journal of Organic Chemistry, *30*, 3037 (1965).
74. Erickson, R. E., Bakalik, D., Richards, C., Scanlon, M., and Huddleston, G., "Mechanism of Ozonation Reactions. II. Aldehydes." Journal of Organic Chemistry, *31*, 461 (1966).
75. Syrov, A. A., and Tsyskovskii, V. K., "Mechanism of the Action of Ozone on Aldehydes." Journal of Organic Chemistry of the USSR, *6*, 1406 (1970).
76. Price, C. C., and Tumolo, A. L., "The Course of Ozonation of Ethers." Journal of the American Chemical Society, *86*, 4691 (1964).
77. Erickson, R. E., Hansen, R. T., and Harkins. J., "Mechanism of Ozonation Reactions. III. Ethers." Journal of the American Chemical Society, *90*, 6777 (1968).
78. Whiting, M. C., Bolt, A. J. N., and Parish, J. H., "The Reaction Between Ozone and Saturated Compounds." Advances in Chemistry Series, No. 77. American Chemical Society, Washington, D.C. (1968), p. 4.
79. Schubert, C. C., S.J., and Pease, R. N., "The Oxidation of Lower Paraffin Hydrocarbons. II. Observations on the Role of Ozone in the Slow Combustion of Isobutane." Journal of the American Chemical Society, *78*, 2044 (1956).
80. Hamilton, G. A., Ribner, B. S., and Hellman, T. M., "The Mechanism of Alkane Oxidation by Ozone." Advances in Chemistry Series, No. 77. American Chemical Society, Washington, D.C. (1968), p. 15.
81. Batterbee, J. E., and Bailey, P. S., "Ozonation of Carbon-Hydrogen Bonds. Anthrone." Journal of Organic Chemistry, *32*, 3899 (1967).

82. Murray, R. W., Lumma, W. C. Jr., and Lin, J. W. P., "Singlet Oxygen Sources in Ozone Chemistry. Decomposition of Oxygen-Rich Intermediates." Journal of the American Chemical Society, *92*, 3205 (1970).
83. Deslongchamps, P., and Moreau, C., "Ozonolysis of Acetals. (1) Ester Syntheses, (2) THP Ether Cleavage, (3) Selective Oxidation of β-Glycoside, (4) Oxidative Removal of Benzylidene and Ethylidene Protecting Groups." Canadian Journal of Chemistry, *49*, 2465 (1971).
84. Mester, L., "Role of Formazan Reaction in Proving Structure of Ozone-Oxidized Carbohydrates." Advances in Chemistry Series, No. 21. American Chemical Society, Washington, D.C. (1959), p. 195.
85. Schuerch, C. "Ozonization of Cellulose and Wood." Journal of Polymer Science: Part C, Polymer Symposia, *2*, 79 (1963).
86. Dobinson, F., and Lawson, G. J., "Chemical Constitution of Coal VI— Optimum Conditions for the Preparation of Sub-humic Acids from Humic Acid by Ozonation." Fuel, *38*, 79 (1959).
87. Dobinson, F., "Ozonation of Malonic Acid in Aqueous Solution." Chemistry and Industry, 853 (1959).
88. Austin, J. D., and Spialter, L., "Substituent Effects in the Oxidative Cleavage of Organosilanes by Ozone." Advances in Chemistry Series, No. 77. American Chemical Society, Washington, D.C. (1968), p. 26.
89. Spialter, L., Pazdernik, L., Berstein, S., Swansiger, W. A., Buell, G. R., and Freeburger, M. E., "The Ozone-Hydrosilane Reaction. A Mechanistic Study." Advances in Chemistry Series, No.112. American Chemical Society, Washington, D.C. (1972).
90. Waters, W. L., Pike, P. E., and Rivera, J. G., "The Ozonolysis of Organomercurials." Advances in Chemistry Series, No. 112. American Chemical Society, Washington, D.C. (1972).

CHAPTER IV

PILOT PLANT STUDIES OF

TERTIARY WASTEWATER TREATMENT WITH OZONE

B. S. Kirk, R. McNabney, and C. S. Wynn

INTRODUCTION

Increasing populations and improving standards of living are placing increasing burdens on water resources. The preservation of our limited natural water supplies and, in the not distant future, the necessity for direct recycling of water in some parts of the country will demand improved technology for the removal of contaminants from wastewater.

The contaminants in wastewater are many and continually varying, and they are not well characterized according to chemical species. Commonly, the level of contamination is described by either a biological oxygen demand (BOD) or a chemical oxygen demand (COD). BOD_5 is a measurement of the oxygen consumption by aerobic microbes feeding on the contaminants during a 5-day incubation. COD is the oxygen equivalent of the oxidation of the contaminants in hot dichromic acid. In general, COD analyses are quicker and more reproducible than BOD_5 analyses. For these reasons, we use COD to characterize the contaminants in wastewater.

Wastewater treatment is usually divided into three stages: primary, the removal of settleable solids; secondary, the removal of readily biodegradable contaminants; and tertiary treatment. Tertiary treatment is generally the further treatment of wastewater after prior treatment has reduced the COD to less than about 60 mg/liter and the BOD_5 to less than about 20 mg/liter. It may also involve the removal of nitrogenous compounds and phosphates.

Tertiary treatment processes include lime (or other chemical) clarification, filtration, activated carbon adsorption, and ozone treatment.   Ozone, O_3, is a very powerful oxidant which can oxidize most of the oxidizable contaminants in wastewater.   It is, however, an unstable material which must be generated at the point of use and in fairly low concentrations.

Ozone has been used for treating drinking water in other countries for many years.   And it has been used to treat some special industrial wastes, notably for removing cyanides and phenols.   However, ozone has not been used for general wastewater treatment on any large scale.

Several years ago, Airco ran experiments on ozone treatment of secondary treated wastewaters.   The favorable results led to a prepilot, bench-scale experimental program sponsored by the Federal Water Pollution Control Administration.

On the basis of these laboratory investigations, it was established that ozone tertiary treatment offers the following potential advantages.

1. There is a major reduction in both BOD and COD. The organic material is destroyed, not simply separated; hence, there are no problems of solids handling or of concentrated wastes disposal.

2. There is pronounced reduction of odor, color, turbidity, and surfactants.

3. All bacteria are killed.

4. The treatment by-products are harmless or beneficial.

5. The product water has a high dissolved oxygen (DO) concentration.   Especially when using pure oxygen to prepare ozone, the DO is generally greater than the residual BOD or COD.   The effect of such effluent on receiving streams would be that of a negative BOD.

Because of the advantages of the process, Airco, under contract to the Environmental Protection Agency, designed, built, and operated a pilot plant for ozone tertiary treatment at Blue Plains, Washington, D.C.

In this paper we will describe the pilot plant, the experimental programs, and some preliminary results.   It must be emphasized that the results are preliminary and are subject to revision.

PILOT PLANT EQUIPMENT

Concept of the Ozone Treatment

The tertiary ozone treatment of wastewater can be conveniently divided into three steps: (1) generation of ozone gas, including preconditioning the feed gas and means of circulating the gas; (2) dissolution of the ozone from the feed gas in the wastewater; and (3) the detention of the ozonated water for a period sufficient for the organic contaminants to be oxidized.

Pilot Plant

A simplified schematic diagram of the experimental pilot plant is shown in Figure 20. Being a research facility, this pilot plant was designed for maximum flexibility and is, therefore, much more complex than a full-size plant would be. Its design nominal flow is 35 gpm (50,000 gal/day) of a pretreated feed having COD of about 60 mg/liter. Utilizing all tankage and every ozone feeder, the nominal treatment time is 1-1/4 hours.

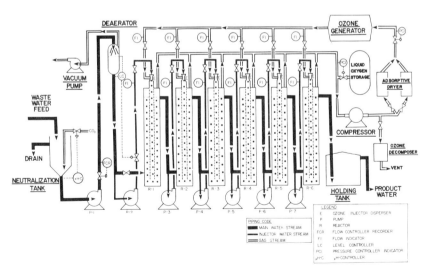

Figure 20. Flow Diagram of the Pilot Plant for Wastewater Treatment with Ozone

Wastewater enters the plant through a surge tank which
has provisions for automatic pH adjustment by a carbon
dioxide sparge system. Dissolved gases, mostly nitrogen,
are removed in a spray-type vacuum deaerator operating at
about 28 in. Hg vacuum. The water then flows through a
series of six 440-gallon reactors, each 24 feet high and
2 feet diameter, in which the actual ozone treatment takes
place. Finally, the treated water is detained in a holding
tank long enough for the residual dissolved ozone to decompose before it is discharged from the plant.

The ozone is produced by electrical discharge in a
commercial ozone generator having a nominal 60 lb/day
capacity. Pure oxygen rather than air is used as the
feed gas in order to realize greater power efficiency in
the generator. In addition, since electrical ozone generators require a bone-dry gas feed for maximum efficiency
and low maintenance, the use of oxygen rather than air
reduces the volume of gas that must be dried.

Although pure oxygen is not very expensive in large
quantities, its cost together with the limited concentrations of ozone (1% to 4% wt) that can be efficiently
produced by existing generators makes recycling of the
oxygen an economic necessity. The aeration of wastewater
with an ozone/oxygen mixture for an hour through the contactors would strip dissolved nitrogen from the water
according to Henry's Law. Since the off-gas is recycled
and reused, nitrogen would build up in the "closed" gas
system and, in time, affect ozone production in the
generator in an undesirable way. The deaeration step
prevents the build-up of inert gases (primarily nitrogen)
in the gas stream. In the Blue Plains plant, the gas
stream usually contains 90 to 95 volume percent oxygen,
the remainder being nitrogen and carbon dioxide roughly
in the ratio 2:1.

The gas flow through the plant cycles as follows: the
gas mixture from the ozonator is fed in parallel to the
dissolving equipment of the six reactors. The spent gas
is collected from the reactors, compressed to about 50
psig in a Nash water-ring compressor, and dried in an
adsorption dryer which is automatically regenerated by
low-pressure purge. Makeup oxygen is added as needed to
maintain a constant suction pressure at the compressor
inlet.

The gas dissolution equipment consists of an ejector-
type, high-shear mixer atop each reactor. The highly
dispersed gas/water mixture is forced down a 4-inch-
diameter dissolver tube to near the bottom of the reactor.

From there, the bubbles rise through the water to the top
of the reactor where the spent gas is collected for
recycle.

Gas and water feeds to each mixer are manually adjusted
and are individually metered. Water feed is supplied by
a separate pump for each mixer. Each pump draws from the
overflow of the preceding reactor and can draw additionally
from the bottom of the reactor it feeds. Through this
arrangement, the water flow through each mixer is independent of the plant feed-water flow rate, and relative
gas and liquid volumes are adjusted to insure optimum
operation of the injectors.

Numerous sampling taps are provided. Gas samples are
available at the inlet and outlet of the ozone generator
and from the gas space at the top of each reactor. Water
samples can be drawn from the surge tank, the deaerator,
the top and bottom of each reactor, and the product tank.

The ozone dosage rate is manually controlled by varying
the gas flow and electrical power input to the ozone generator. Coarse adjustment is made by proportioning the
ozone dosage to the feed-water flow, and finer adjustment
is made to maintain the desired dissolved ozone concentration at the outlet of each reactor.

Pilot Plant Materials

The surge tank, reactors, and holding tank are made of
mild steel. The reactors have a hot-applied bitumastic
lining, and the holding tank is coated with solvent-applied
bitumastic. Pipes carrying dry gas are aluminum, and those
carrying the wet spent gas are stainless steel. Water
piping and the mixers are mostly polyvinyl chloride (PVC),
with some stainless steel fittings. All pumps and the
compressor are made of stainless steel. After nine months
of operation, no corrosion problems have become evident
with any of these materials.

Post-Ozonation Experiment

The apparatus shown in Figure 21 has been used to gather
additional data for the ozonation process. Product water
is placed in the 3-liter flask and subjected to ozone
sparging for an additional 4 to 8 hours in order to carry
the COD reduction nearly to completion. The gas sparge is
then stopped; contemporaneously, the liquid and gas are
analyzed for ozone concentration and a sample for COD
analysis is taken. Then over the following 40- to 60-minute
period, several analyses of the dissolved ozone are made.

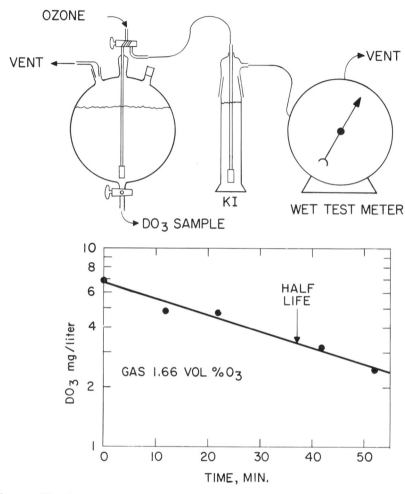

Figure 21. Post-ozonation Experiment for Determining End
Point on COD Reduction Curve, Solubility of
Ozone, and Auto-decomposition Rate of Dissolved
Ozone

By means of this extended ozonation procedure, three
data facts are attained. The COD analysis affords an end
point on the reaction curve. From the contemporaneous
gas and liquid analyses, ozone solubility in the wastewater

can be calculated. And, by plotting the log of dissolved ozone concentration against time, as is shown in the bottom of Figure 21, the rate of ozone autodecomposition in the wastewater (assuming first order kinetics) can be determined.

PLANT LOCATION

The pilot plant is located at the Blue Plains sewage treatment plant in Washington, D.C., a large plant serving the Washington metropolitan area. The wastewater is nearly all domestic waste and, therefore, has a fairly low level of COD and BOD. The feature of this location is the presence of a complex of pilot scale, advanced wastewater treatment processes operated cooperatively by the Environmental Protection Agency and the District of Columbia. These facilities afford a large variety of feeds to the ozone treatment pilot plant.

A list of pilot plant scale pretreatments available at Blue Plains is presented in Table 5, together with the abbreviations assigned to each. The examination of all possible combinations of secondary and tertiary pretreatment is impractical. The feeds that have been treated in the ozone plant are listed in Table 6, together with the average COD of each feed. The feeds have ranged from those typical of many secondary effluents to those so well pretreated that ozone treatment is only a polishing step, yielding sterile water suitable for direct recycling. It may be noted that only the straight IPC feed has a COD near the plant design parameter.

PLANT OPERATION

Normally, the ozone pilot plant is operated 24 hours a day, 5 days a week. Plant operating parameters are monitored continuously, and, once during each 8-hour shift, a complete set of plant data is obtained from samples collected on a time-delay sequence. These data ostensibly trace the flow of a slug of wastewater through the plant. Data collected include all gas and liquid flows and pressures; the COD as the water slug leaves each reactor; the ozone concentration in the gas from the ozone generator and in the spent gas from each reactor; the pH, temperature, and turbidity of the feed and product water; the dissolved ozone concentration in each reactor; the ozone generator

Table 5. Sources of Wastewater at Blue Plains

I. E.P.A./D.C. Pilot Plant

   A. Independent physical/chemical treatment

     IPC   —Lime (and mineral) clarified raw wastewater

   B. Biological treatment

     UNOX —Oxygen activated sludge

     3X-BIO—Three-stage biological treatment

          (1) High-rate activated sludge
          (2) Aerobic nitrification
          (3) Anaerobic denitrification

   C. Tertiary physical/chemical treatment

     LIME —Lime clarification

     FILT —Dual-media filtration

     CL   —Break-point chlorination for ammonia removal

     C   —Activated carbon adsorption

II. Blue Plains Plant Effluent

Table 6. Wastewater Streams Treated in Ozone Pilot Plant

Pretreatment	COD, mg/liter
IPC	54
IPC + CL	43
IPC + CL + C	15
UNOX	24
UNOX + LIME + FILT	17
3X-BIO + FILT	15
3X-BIO + LIME + FILT	22

power parameters; and the makeup oxygen consumption. These data are entered onto computer data sheets by the plant operators for later computer reduction of the data.

A consideration worth noting is that the ozone pilot plant responds very rapidly to changes in operating parameters. The plant can be started up in less than 20 minutes and closed down in less than 10 minutes. Because of the rapid plant response, composite sampling is unnecessary.

PRELIMINARY RESULTS

Reaction Rate

In Figure 22 are typical COD reduction curves for two

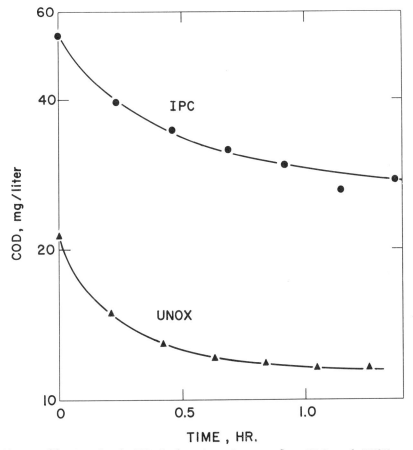

Figure 22. Typical COD Reduction Curves for IPC and UNOX
Pretreated Feeds: Log COD versus Reaction Time

plant feeds in which log COD is plotted versus reaction
time with ozone. Two characteristics are evident. The
initial reaction is quite rapid, but, as the reaction
continues, it slows down so drastically that one may
suspect that it has neared completion.

In Figure 23 log COD (normalized to feed COD) is plotted
against log reaction time. These data are from a number
of runs on UNOX effluent and include end-points from the
post-ozonation experiment. The convex shape of these
curves demonstrates that practically all of the COD can
eventually be oxidized by ozone. Thus, in the technical

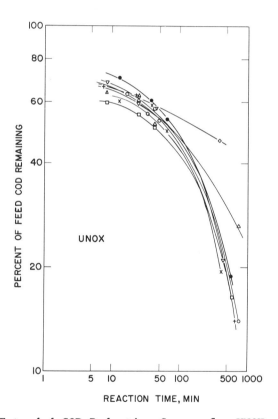

Figure 23. Extended COD Reduction Curves for UNOX
           Pretreated Feeds including End-Points from
           Post-Ozonation Experiment: Log COD versus Log
           Reaction Time

sense, there appears to be no fraction of the COD that is refractory to ozone treatment. However, the reaction time becomes so extended after 50 to 70% reduction that the economic practicality of further reduction becomes questionable.

pH Effects

The pH of the wastewater is an obvious process parameter, and considerable time has been spent examining its effects. Somewhat surprisingly, the pH of the wastewater consistently changes toward neutrality during ozone treatment. In Figures 24, 25, and 26 are shown the pH changes of many runs on several different plant feeds. The circular symbols representing the feed-water pH are connected by vertical lines to the triangular symbols representing the product water pH. Whether the feed water is acidic or basic, the product water is always nearer neutrality. Limited investigations show that most of this pH change takes place in the first part of the reaction. It also appears that the pH change is greater for the higher COD feeds.

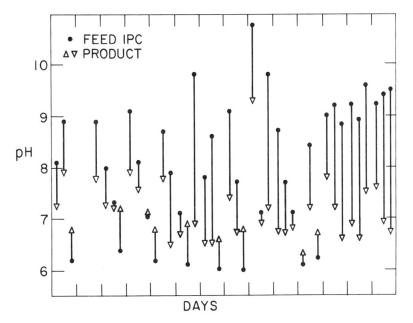

Figure 24. Change of pH during Ozone Treatment of IPC Pretreated Feed

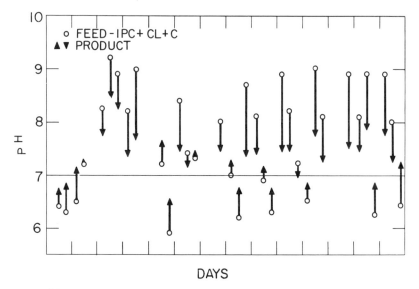

Figure 25. Change of pH during Ozone Treatment of IPC +
CL + C Pretreated Feed

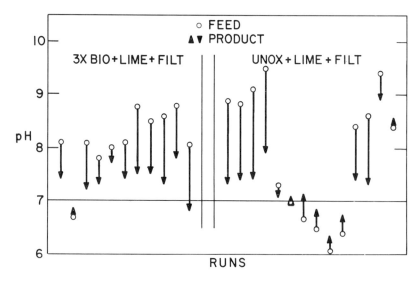

Figure 26. Changes of pH during Ozone Treatment of 3X–BIO +
LIME + FILT and UNOX + LIME + FILT Pretreated Feed

There are no provisions for controlling the pH throughout the reaction period. Therefore, in examining the
effects of pH on COD reduction, we can use only the feed-
water pH as an independent variable.

Recognizing this constraint, the effects of feed-water
pH on COD reduction are summarized in Figure 27. Each
point on this graph represents the average of a number of
runs. These data show that the higher the pH of the feed,
the higher the COD removals realized in one hour's ozone
treatment. And, generally, the pH effects are more pronounced on the biologically pretreated feeds (UNOX and
3B) than on the physicochemical pretreated feeds (IPC).

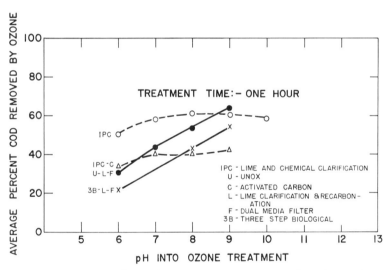

Figure 27. Effect of Feed-Water pH on COD Reductions in
One Hour for Four Different Pretreatments

Reaction Rate Constants

In the computer reduction of the pilot plant data, a
number of reaction rate constants are evaluated. At this
point, however, the correct kinetic expression which best
describes the process has yet to be defined.

Generally, both first order (Equation 1) and second
order (Equation 2) kinetics are considered.

$$dL/dt = -K_1L \tag{1}$$

$$dL/dt = -K_2LZ \tag{2}$$

where
   L is COD, mg/liter
   t is time, hours
   Z is dissolved ozone, mg/liter

These two equations can be integrated in several ways for each reactor. Both can be integrated assuming plug flow or complete mixing in the reactor, but, with multiple reactors, the final results differ very little. And there are several ways in which the dissolved ozone concentration in Equation 2 can be evaluated.

For this discussion, only two reaction rate constants will be considered: a first order constant, K_1, obtained from Equation 1 assuming plug flow through the reactor, and a second order constant, K_2, obtained from Equation 2 assuming perfect mixing in each reactor and taking for the value of Z the average of the dissolved ozone concentration measured at the top and bottom of the reactor.

Figure 28 shows a semilog plot of K_1 and K_2 against the percent of feed COD remaining. The points were calculated from data for the IPC pretreated feed with pH between 8.7 and 9.2 and using COD values which have been analytically smoothed to reflect a continuous decrease through the chain of reactors. The data show some scatter as wastewater treatment data always do, and K_1 and K_2 show about the same degree of scatter. However, both reaction rate constants clearly fall near a straight line, and both decrease by a factor of about ten by the time about 60% of the COD has been removed. This gross change in reaction rate constants reflects the spectrum of reactivities of the wastewater organics with ozone.

That the analytical smoothing of the COD data is not creating this behavior is shown in Figure 29. The coordinates for this plot and the influent are the same as those used in Figure 28, but calculated from the raw, unsmoothed COD data. Although the raw data naturally exhibit more scatter, the linearity of the trend is still evident.

Ozone Consumption

In the ozone oxidation of organic materials, it is generally expected that only one atom of oxygen from the

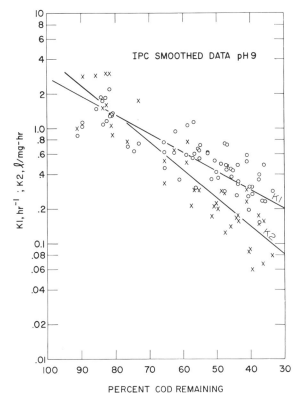

Figure 28. Variation of First Order (K_1) and Second Order
(K_2) Reaction Rate Constants Calculated from
Analytically Smoothed COD Data for IPC Pre-
treated Feed at pH 9.0 ± 0.3

O_3 molecule is highly reactive. On this basis, the oxidation of one pound of COD should consume three pounds
of ozone and yield two pounds of molecular oxygen as a
by-product.

In Table 7, we show the ozone consumption, both as mg
O_3 consumed per liter feed water and as lb O_3 used per
pound COD removed during the treatment of IPC-pretreated
feed water. These data are the average of 33 sets of
data in which only five reactors were in use. The pH of
the feed water ranged from 6.0 to 9.8.

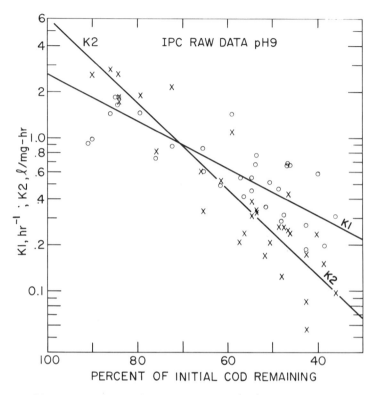

Figure 29. Variation of First Order (K_1) and Second Order
(K_2) Reaction Rate Constants Calculated from
Raw COD Data for IPC Pretreated Feed at pH
9.0 ± 0.3

Three facts are learned from these data. First, the
absolute ozone consumption, mg/liter, is greater in the
first stages since most of the COD reduction takes place
there. Second, the lb O_3/lb COD removed increases rapidly
as the degree of treatment increases, which indicates the
increased resistance to ozone oxidation of the remaining
organics. And third, in the first stage and even for the
overall treatment through five reactors, less than the
expected 3.0 lb O_3/lb COD removed are consumed, which
suggests that auto-oxidation figures into the reaction
mechanism.

Table 7. Average Ozone Consumption, IPC Feed (33 Operating Shifts)

pH Range: 6.0-9.8
Feed COD: 54 mg/liter
Ozone Doses: 2.47 lb O_3/lb COD Feed
Time per Stage: 12.5 - 22 min

Stage No.	O_3 Used (mg/1)	O_3 Used (lb/lb COD removed)
1	31.8	2.15
2	18.3	3.11
3	14.3	3.42
4	9.4	4.05
5	8.4	5.04
Overall	82.2	2.70

Nitrogenous Materials

In addition to COD removal, the removal of nitrogenous materials from wastewater is important. Most of the nitrogen in wastewater is ammonia, which is not measured in the COD determination. If the nitrogen compounds cannot be completely removed (as nitrogen gas, for example), it is desirable to eliminate subsequent oxygen demand by at least oxidizing them to nitrates.

Our investigation of nitrogen removal has been limited in scope, and all of the nitrogen analyses have been done for us by the EPA analytical group at Blue Plains.

A summary of the nitrification of nitrogenous compounds in the IPC plant feed is shown as a bar chart in Figure 30. Each bar is the average of several runs. For three feed-water pH's, we show the nitrogen analyses of the feed water, the effluent from the first reactor, and the product water leaving the sixth reactor. Values are given for total Kjeldahl nitrogen (TKN), ammonia nitrogen (NH_3-N), and nitrate nitrogen (NO_3-N).

These data indicate that nitrification during one-hour ozone treatment is significant only with high pH feeds.

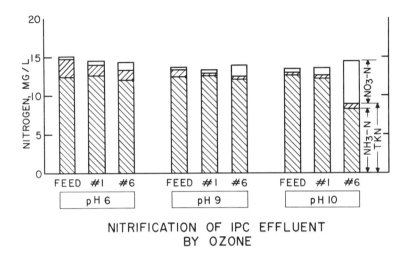

NITRIFICATION OF IPC EFFLUENT
BY OZONE

Figure 30. Oxidation of Nitrogenous Compounds by Ozone

Interestingly, no detectable removal of ammonia occurred
in the vacuum deaerator, even at high pH.

Turbidity

The reduction in turbidity, expressed as Jackson
Turbidity Units (JTU), during ozone treatment is illustrated by the data in Figure 31 for two feed waters,
neither of which had been filtered. In both cases,
turbidity reductions of about 70% were realized. The
product water ranged generally between 1 and 4 JTU, which
is reasonably clear water. On filtered feeds, turbidities
less than 1 JTU were generally obtained.

Disinfection

Periodically, water samples were taken for bacteriological analysis by an independent laboratory. The
results of these analyses for samples of feed water and
the effluent from the first, third, and sixth reactors
are summarized in Table 8. Practically all bacteria were
killed in the first reactor, and none survived the entire
treatment.

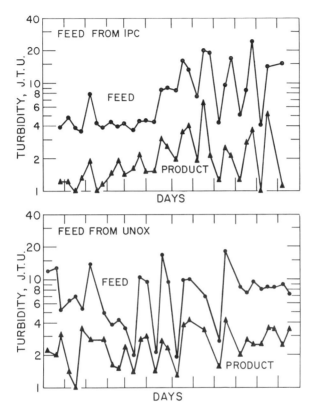

Figure 31. Turbidity Reduction During Ozone Treatment for
IPC and UNOX Pretreated Feeds

Electrical Power Requirements

Figure 32 shows some illustrative data for the electrical power requirement for the process which is the
major part of the operating costs. For this illustration,
we arbitrarily selected all of the runs made on IPC feed
water in the pH range 8.0 to 9.4. This feed is the
strongest feed water (highest COD) we have treated and,
for this reason, reflects the maximum power requirements
per 1000 gallons we have observed. It is well to recall

Table 8. Disinfection of Wastewater during Tertiary
         Treatment with Ozone

Source of Feed	Feed	#1	#3	#6
Standard plant count/ml				
UNOX + LIME + FILT	12,000	-0-	-0-	-0-
IPC + CL + C	110,000	-0-	-0-	-0-
IPC	150,000	7	-0-	-0-
3X-BIO + LIME + FILT	2,000	-0-	-0-	-2-
Most probable number coliform/100ml				
UNOX + LIME + FILT	23	<3	<3	<3
IPC + CL + C	3.6	<3	<3	<3
IPC	<1100	<3	<3	<3
3X-BIO + LIME + FILT	46	0	0	0
Most probable number *E. Coli*/100ml				
UNOX + LIME + FILT	<3	<3	<3	<3
IPC + CL + C	<3	<3	<3	<3
IPC	<3	<3	<3	<3
3X-BIO + LIME + FILT	2.3	0	0	0

that the plant was designed as an experimental unit, and
that the complexity of equipment and redundancy of units
facilitated flexibility rather than economy of design or
operation.

Each bar represents a single run. The power consumption
is subdivided as indicated on the right side of the figure.
The ozone generator power reflects the total ozone produced, not merely the amount used in reaction, and also
includes the ozone found in the off-gas due to incomplete
dissolution. (In these runs, about 70% of the ozone feed
was dissolved.) Since the pilot plant equipment is in-
tentionally oversized, pump and compressor work are calculated from flow rates and pressure drops assuming 70%
mechanical efficiency.

Although not occurring in this sequence, the compressor
power is computationally divided into two parts. The power
for gas recycle is the power required to compress the
recycle gas from reactor pressure (about 2 psig) to the
ozone generator pressure (about 12 psig). Gas drying

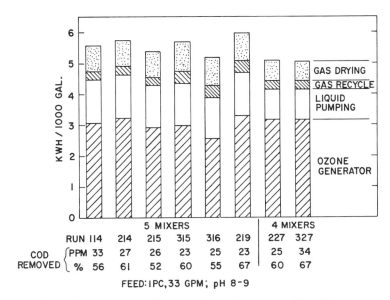

Figure 32. Illustrative Energy Requirements for Ozone
Tertiary Treatment IPC Pretreated Feed at pH
8 to 9

power is that additional compressor work required to
increase the gas from ozone generator pressure to that of
the dryer (about 50 psig).

For this comparatively high COD feed, we see that the
power requirements are between 5 and 6 Kwh/1000 gal,
roughly 3 Kwh/1000 gal being consumed in the ozone generator and 2 to 3 Kwh/1000 gal being consumed by pumps
and compressors.

Along the bottom of Figure 32 are data for each run.
The first six runs were made using five of the six reactors
and the last two runs using only four of the six reactors.

SUMMARY

All of the results presented here are preliminary and
are subject to revision. Because of the interim nature
of the material presented, it is inappropriate to present
firm conclusions since they, likewise, could be subject to
revision. Instead, a simple summary and a present estimate
of this study is given.

A pilot plant for the tertiary treatment of wastewater with ozone has been described here. It has been operated on feed water subjected to a variety of secondary and tertiary pretreatments.

The COD reduction rate is rapid in the first stages of treatment but slows down drastically in the latter stages after 50-70% of the COD has been removed. However, if treatment is continued long enough, practically all of the COD can be oxidized by ozone.

The pH of the wastewater changes toward neutrality during ozone treatment. Generally the higher pH feed waters show a faster COD reduction.

At least for the IPC pretreated feed water, the ozone consumption in the first stage is less than the expected 3 lb O_3/lb COD removed. In the latter stages, the ozone consumption is substantially greater. However, because most of the COD is removed in the first stage, the average ozone consumption is still less than 3 lb O_3/lb COD removed.

For the limited cases correlated so far, the log of both the first order and second order reaction rate constants appear linear with percent COD removal, and both decrease by roughly a factor of ten during treatment.

The oxidation of nitrogenous material to nitrates during ozone treatment is significant only at high pH.

Major reductions of turbidity have been realized on all plant feeds, and the ozone tertiary treatment, as expected, produces complete bacteria kills.

It is evident that pump and compressor power requirements are not insignificant compared to ozone generation power requirements.

In overall summary, the pilot plant has generally met design expectation in its operations.

CREDIT

The work described in this paper was done under contract for the Environmental Protection Agency, EPA Project No. 17020-DYC and Contract No. 14-12-597.

CHAPTER V

OZONE AS A WATER AND WASTEWATER DISINFECTANT:

A LITERATURE REVIEW

Albert D. Venosa

INTRODUCTION

Disinfection is the final polishing step in the purification of water and wastewater. Without it, water supplies would be unsafe for commercial use. Until now, chlorination has been the principal mode of disinfection of water and wastewater in the United States. Concern has been expressed recently regarding the possible formation of toxic chlorinated end products resulting from chlorination of both domestic and industrial waste discharges. Evidence has been accumulating to suggest that chlorinated effluents adversely affect aquatic life.[1] This has led some workers to search for other methods of disinfection that would result in little or no toxicity to aquatic life, and ozone is a disinfectant that is being considered as a possible alternative to chlorine.

Ozone, an allotropic form of oxygen, is a powerful oxidizing agent, the oxidation-reduction potential being +2.07 volts as compared with that of HOCl which is +1.49 volts (at 25° C and unit hydrogen ion activity). Ozone is more soluble in aqueous solution than oxygen but because of a much lower partial pressure it is difficult to obtain more than a few milligrams per liter concentration under normal conditions of temperature and pressure.[2] Ozone is unstable in water, having an effective half-life measured in minutes. Its decomposition is accelerated by neutral salts and hydroxyl ions. Thus, it must be produced on the site where it is to be used.

Of the various means by which ozone can be generated,
electrical production is the only practical and economical
method for large scale use.[3] It is produced by passing
clean, dry air or oxygen between electrodes across which
is maintained an alternating high-voltage potential.

The principal ozone decomposition products in aqueous
solution are molecular oxygen and the highly reactive free
radicals HO_2, OH, and H.[4] Very little is known about the
significance of the free radical intermediates on the
germicidal properties of ozone solutions. The same free
radicals are produced by irradiation of water,[5] and it
has been reported that HO_2 and OH radicals contribute
significantly to the killing of bacteria by irradiation.[6]

SURVEY OF THE LITERATURE

It is uncertain when the bactericidal properties of
ozone were first recognized. In 1873, Fox[7] stated that
experiments with fluids containing organic matter in a
state of decomposition have shown that ozone destroyed
the most elementary forms of life, such as mold, fungi,
and bacteria. An article published in Engineering News
in 1910[8] reported that the utility of ozone as a bactericide through its marked oxidizing properties was demonstrated by de Meritens in 1886 and Froelich in 1890. In
an article discussing the history of chlorination of
drinking water,[9] it was stated that the first attempt to
sterilize water by chemical means was made in 1893 by
Ohlmueller, who conducted experiments on the action of
ozone on bacteria. Ohlmueller had found that ozone could
not be economically used in treating water containing much
organic matter because of the high demand exerted by the
oxidizable compounds. Thus, the method gained little
popularity at the time.

In a review article by Blanke in 1928,[10] it was pointed
out that Tindall aroused the interest of the scientific
community when he demonstrated the practicability of
sterilizing water with ozone.

Early attempts to use ozone in water treatment were
confined almost exclusively to sterilization, and little
or no thought was given to its ability to oxidize organic
matter, though this property was known from the first uses
of ozone.[11] There was much in the early technical literature on the bactericidal action of ozone, but only occa-
sional mention was made concerning its ability to eliminate
objectionable tastes and odors.[11]

Several experimental ozone plants were in operation as
early as 1892. Van Ermengen[12] and later Calmette and
Roux[13] found that ozone killed all pathogenic and sapro-
phytic microbes encountered in water, even the most
resistant sporulating types of bacteria. The first really
important application of ozone in water treatment was
operational in Nice, France, in 1906.[2]

In 1933, it was voted unanimously in Paris by the
Scientific Commission for the Study and Supervision of
Water Purification that ozonation was the best method for
water sterilization.[14] Later, de Lipkowski,[15] in justify-
ing the widespread use of ozone in France, reported the
virtual disappearance of typhoid fever and other water-
borne diseases in those cities where the water supply was
disinfected with ozone. Lebout[16] concluded that of 92
cases of typhoid and paratyphoid fever occurring in Nice
in 1948, none were attributable to transmission by the
municipal drinking water which was being treated with
ozone.

In the United States, the use of ozone has been primarily for taste, odor, and color control; therefore,
bacteriological data from American sources are sparse.
In 1935 Ferkinoff[17] reported the removal of 99.2% of
Escherichia coli from filtered raw river water, and 100%
removal from ozonated water. Four years later, Cox[18]
stated that it was possible to disinfect filtered water
adequately when sufficient ozone was added.

In July 1940, the city of Whiting, Indiana, placed
ozone generating equipment into service because the industrial development and commensurate increase in water
pollution had caused a crisis with respect to taste- and
odor-producing industrial waste pollutants in the raw
water supply.[19] Ozonation was used, not for disinfection
purposes, but as a pretreatment before chlorination.
Nevertheless, some bacteriological data were collected by
Bartuska, who reported an average reduction of 95 to 97%
in coliform index.[19]

In 1941, Consoer[20] reported favorable results from
several American ozonizing plants. At Long Beach, Indiana,
the water purification plant located near a primary sewage
outfall had produced a satisfactory finished water for 10
years. The same was reported for the Hobart, Indiana;
Whiting, Indiana; and Denver, Pennsylvania, plants. The
total annual cost for ozonizing the Whiting water was
$4.44 per million gallons.

In 1949, the city of Philadelphia built the world's
largest ozone plant for removal of tastes, odors, and

manganese from Schuykill River water.[21]   Coliform inactivation was proportioned to ozone residual.   The kills were
approximately 90.5% for average ozone residual of 0.06
mg/liter, 94.5% for 0.08 mg/liter, and 98% for 0.16 to
0.20 mg/liter.

In 1944, Smith and Bodkin[22] compared the bactericidal
action of ozone and chlorine at varying values of pH.
Ozone, over a wide pH range, was many times as effective
as chlorine.   At a temperature of 27.5° C and pH 5.0 and
6.0, ozone effected sterility of a 1-liter sample containing 8×10^5 bacteria/ml in 5 minutes.   At pH 7.0, 8.0,
and 9.0, the sterilization time was 7.5 minutes.   The
ozone concentration varied from 0.13 to 0.2 mg/liter.   In
contrast, the concentration of chlorine required to
sterilize as rapidly as the ozone varied from 2.7 mg/liter
at pH 5.0 to 7.9 mg/liter at pH 8.0.   It was not stated
whether the chlorine was in the free or combined state,
and the characteristics of the test water used were not
described.   In 1946, Yanshina[23] stated that an ozone
residual of at least 0.5 mg/liter in the final effluent
was necessary for adequate disinfection of water supplies.

Howlett[24] discussed the application of ozone to water
purification and pointed out that although ozonation was
more costly than chlorination, possibilities existed for
cost reduction.   One year later, in 1948, Whitson[25]
presented in detail a comparison between disinfection
with ozone and with chlorine, basing his observations on
an ozone plant installed in 1936 with a capacity of 10
mgd.   Although more expensive, the ozonized water had a
better appearance and exhibited no ozone residual.   Further, no deterioration in bacteriological quality was
observed when the sterile ozonized water was passed into
the distribution system.

In 1949, a detailed account of the bactericidal action
of ozone in both pure water and water from the River Plate
was presented by Leiguarda, et al.[26]   Varying amounts of
ozone were added to pure water free from ozone demand,
and the water was then inoculated with dilute suspensions
of *E. coli* or *Clostridium perfringens*.   Samples were taken
at intervals to determine the concentration of ozone and
the number of bacteria present.   The effects of temperature
and pH on bactericidal action were also investigated.
Results indicated that in water initially containing 10^4
E. coli/ml and 0.12 mg/liter ozone, at pH 6.0 and maintained at a temperature of 10° C, no viable bacteria were
found after 5 minutes; the ozone content had decreased to
0.09 mg/liter.   At pH 8.0 and higher temperatures, the

ozone concentration was reduced more rapidly; bactericidal
efficiency was not significantly affected by temperature
but was slightly greater at pH 6.0 than at pH 8.0. In
addition, tests were made on the effects of ozone on the
naturally occurring bacterial flora of water from the
River Plate. The ozone demand of this water was high,
and tests were made using 1 to 6 mg/liter of ozone. There
were sizable reductions in the number of bacteria even
when the amount of ozone added was insufficient to satisfy
the ozone demand of the water. All organisms were destroyed
when, after a contact period of 5 minutes, 0.08 mg/liter of
residual ozone was present. Similar results were obtained
in experiments with river water which was coagulated,
settled, and subsequently inoculated with *E. coli*. A total
kill of vegetative forms of *C. perfringens* occurred within
5 minutes when 0.12 mg/liter ozone was initially added to
water containing 1.4×10^4 bacteria/ml. In water containing *C. perfringens* in concentrations of 2×10^3 spores/ml,
at a pH of 6.0 and maintained at a temperature of 24° C,
no viable spores were found after a contact period of 15
minutes with 0.25 mg/liter of ozone, or after 2 minutes
with 5 mg/liter. At pH 8.0, bactericidal efficiency was
reduced; spores were not affected by 0.25 mg/liter of
ozone even after 120 minutes.

Ingram and Haines[27] reported that the susceptibility of
fungi and bacteria to ozone in water was decreased by at
least one order of magnitude when the organisms were exposed to the disinfectant in nutrient broth.

Cysts of *Endamoeba histolytica*, which were relatively
resistant to chlorination, were easily inactivated by
ozone in aqueous solutions.[28] This cysticidal action did
not appear to be influenced by pH, temperature, or organic
nitrogen content of the water. A 99% kill of the cysts
in 1 to 3 minutes was observed in an aqueous solution
containing 0.5 to 1.0 mg/liter ozone.

Novel and Buffle[29] conducted a comparative study of
the sterilizing action of chlorine and ozone on the water
of Lake Geneva in the Petit-Lac. Ozone, at a concentration
of 1 mg/liter, reduced the bacterial content of filtered
lake water from 190/ml to less than 1/ml in one minute;
chlorine, at the same concentration, reduced the viable
numbers to only 40/ml in 5 minutes and to 2/ml in 40
minutes.

In a 1950 symposium on the sterilization of water,
Whitson[30] reported that ozonation effected microorganism
removal, better filterability, and improved color, taste,
and odor. In the same year, Buffle,[31] while comparing the

bactericidal efficiencies of chlorine and ozone, asserted
that ozone was the far superior disinfectant, being considerably faster than chlorine and not as notably affected
by external factors such as pH and temperature.

The water supplies of Nice and the nearby coastal communities were being treated with ozone following filtration in 1950.[32]   The dosage of ozone needed for sterility
of the filtered river water was 1 mg/liter for a contact
time of 5 to 20 minutes.

In 1952, an ozone pilot plant was built in Iowa City in
an attempt to resolve the highly objectionable taste and
odor problems of the Iowa River water.[33]   The coliform MPN
of raw river water ranged from 160,000/100 ml to 400/100
ml, while that of ozonated water ranged from 900/100 ml
to less than 1/100 ml.   The average percent reduction in
coliform numbers was thus reported to be 99.4%, in close
agreement with the 95% reduction reported by Bartuska.[19]

In 1953, Gubelmann and Scheller[34] conducted experiments
to determine the amount of ozone required to insure a
bacteriologically safe water.   The results pointed out the
important effect of water quality: the most unfavorable
results occurred when the raw water contained sludge.   The
average amount of ozone needed to disinfect filtered water
was calculated at 0.5 to 0.6 mg/liter.   Rohrer[35] also
found that the presence of readily oxidizable organic
matter increased the amount of ozone required for disinfection of water.

Hettche and Ehlbeck,[36] studying the inactivation of
poliomyelitis virus with chlorine dioxide and ozone,
found the following residuals were sufficient for disinfection of water: ozone, 0.15 mg/liter; ClO_2, 0.08
mg/liter; chlorine, 0.25 mg/liter.

Dickerman, et al.[37] studied the effect of ozone on
bacteria in both tap water and raw stream water at 28° C.
The organisms tested were staphylococci, enterobacteria,
bacilli, and pseudomonads.   In every instance, except with
Bacillus subtitis, application of ozone to produce a
residual concentration of 2 mg/liter was sufficient to
kill all the bacteria in one minute; ozone concentrations
were determined by starch-iodide titration.   In raw water
samples rich in organic matter, higher concentrations of
ozone were required.   It was concluded that a water
treatment installation would have to maintain an ozone
residual of 2 mg/liter for at least 5 minutes to disinfect
raw water low in organic matter.

Holluta and Unger[38] supported Dickerman et al.[37] in
that the presence in the water of substances combining

with ozone must be taken into account in determining the
amount of ozone used. However, figures lower than those
of Dickerman's for bacteriological disinfection were reported. In unpolluted waters ozone concentrations of 0.1
to 0.2 mg/liter after 5 minutes exposure insured sterility;
only with very polluted water or with water having abnor-
mally high bacterial counts were amounts of 0.4 to 0.5
mg/liter required. It should be noted that these investigators added the bacterial suspensions to previously
ozonized tap water rather than ozonizing the bacterial
suspension directly.

Bringman[39] observed that 0.1 mg/liter of active chlorine
required 4 hours to kill 6 x 10^4 *E. coli* cells in water,
whereas 0.1 mg/liter of ozone required only 5 seconds.
When the temperature was raised from 22° C to 37° C, the
ozone inactivation time decreased from 5 seconds to 0.5
seconds. A very careful investigation of the kinetics
of ozone disinfection has been made by Wuhrmann and
Meyrath.[40] During each experiment, the ozone concentration was kept constant by continuously bubbling air
containing ozone through the test solutions. The results
indicated that ozone disinfection was mainly a function
of contact time, ozone concentration, and water temperature.
These investigations revealed that the contact time with
ozone necessary for 99% destruction of *E. coli* was only
one-seventh that observed with the same concentration of
hypochlorous acid. The death rate for spores of *Bacillus*
species was about 300 times greater with ozone than with
chlorine.

Similarly, Kessel et al.[41] found ozone to be many times
more effective than chlorine in inactivating the poliomyelitis virus. Identical dilutions of the same strain
and pool of virus, when exposed to 0.5 to 1.0 mg/liter of
chlorine and 0.05 to 0.45 mg/liter of ozone, were de-
vitalized within 1.5 to 2 hours by chlorine, while only
2 minutes were required with ozone. However, Ridenour
and Ingols[42] previously demonstrated that the difference
in bactericidal activity between ozone and free chlorine
in water was much less than that reported by Kessel and
co-workers.

Lagrange and Rayet[43] compared the efficacy of ozone
and chlorine for the destruction of the parasite
Schistosoma mansoni in water. Sterilization was accomplished in 4 minutes with 1 mg/liter chlorine and in
3 minutes with 0.9 mg/liter ozone.

In 1956, Hann[44] presented a detailed review of the
differences between chlorination and ozonation as

determined by other workers. The latter method of disinfection was found to be somewhat more expensive. Turbidity
interfered with its use, and organic demand had to be
satisfied before germicidal action was effective. Regarding power consumption and costs of ozonation, the author
made the following comments: "Because ozone plants are
electrically operated, almost the entire operating cost
is electrical. For each pound of ozone produced and
applied, 10 kwhr are consumed in large water plants, and
perhaps 12-15 kwhr in small ones. Thus, for an ozone
dosage of 1 ppm, the electrical cost might range from 75
cents to $2 per 1 mil. gal. of water treated, depending
on plant size and electrical-energy rates. . . . Ozone
plants need relatively little operating supervision or
maintenance, and no added personnel are required for these
services. A generating plant to produce and apply the
ozone, however, represents a substantial capital invest-
ment. Amortized over the life of the equipment, which is
approximately 25 years, this fixed charge will generally
amount to about as much as the operating cost."

Fetner and Ingols[45] tested the germicidal action of
ozone and of chlorine against *E. coli* at 1° C in distilled
water. The lethal ozone concentration was 0.4 to 0.5
mg/liter and was independent of contact time greater than
1 minute. Chlorine, at a concentration of 0.25 to 0.30
mg/liter, effected the same kill in 1 to 10 minutes.
Whereas the bactericidal activity of chlorine increased
with concentration and contact time, ozone was ineffective
below its critical concentration (0.4 to 0.5 mg/liter), an
exemplification of the reported "all-or-none" effect.
These investigators presented a symposium in 1957 on the
use of ozone in water treatment,[46] with special emphasis
on cold water. Aside from disinfection data, a simple,
rapid technique for measuring ozone concentration in
aqueous solution was recommended. Following their pre-
sentation, Dr. A. T. Palin stated in the discussion period:
"the general practice in the literature was to quote
figures for ozone doses calculated on the assumption that
all the ozone in the applied ozonized air was transferred
to the water. . . , it seemed, nonetheless, to be important to stress for the benefit of the ozone plant operator
that applied doses would need to be very much higher,
possibly twice as great, if anything like the same concentrations of dissolved ozone were to be achieved." It
was further agreed in the discussion period that measuring
ozone iodometrically at acid pH's resulted in gross errors
in ozone quantitation because of physical chemical factors
not fully understood.

Stumm[47],[48] investigated some of the chemical aspects of ozonation of water. Ozone decomposed more rapidly in water than in air, forming a complicated series of short-lived intermediates, all of which were oxidizing agents and bactericides. In natural waters, organic and inorganic materials were oxidized by ozone, forming compounds called ozonides. Many or all of these were assumed to exert bactericidal effects. Stumm concluded from the information reported that ozone should be considered as a potential disinfectant for water and sewage, especially with respect to the destruction of those pathogenic organisms (such as cysts, viruses, and spores) that are highly resistant toward the action of chlorine compounds.

A research group at the Armour Research Foundation, Chicago, Illinois, and the Biological Warfare Laboratories, Fort Detrick, Maryland,[49] investigated the possibilities of disinfection and sterilization of sewage with ozone. The treatment of sewage effluent was studied by bubbling ozone through raw or autoclaved sewage which had been seeded with tracer organisms. The laboratory results indicated that ozone could be successfully used not only for disinfection but even for sterilization of sewage containing *Bacillus anthracis,* influenza virus, and *B. subtilis* morph. *globigii,* and for inactivation of toxin of *Clostridium botulinum.* Sterility of the seeded raw sewage was attained in most experiments after an ozone consumption of 100 to 200 mg/liter for 30 minutes. Inasmuch as the objective of the investigation was to evaluate the bacteriological effectiveness of the system, no attempt was made to improve ozone efficiency beyond that necessary for the success of the sterility experiments.

In 1957, Guinvarc'h[50] presented results from a three-year study of the Saint Maur ozone plant in Paris. Following sand filtration and ferric chloride coagulation, the water was ozonated to a concentration of 0.1 mg/liter at temperatures below 10° C, and to 0.05 mg/liter at temperatures above 10° C. The addition of ozone was controlled automatically by measuring the change in electrical potential caused by the presence of ozone. After treatment, the water was free of *E. coli,* and the numbers of *C. perfringens* were reduced by 50%.

Army investigators[51] found that the process utilizing gaseous ozone in contact with contaminated sewage proved to be the most economical and effective method for disinfecting sewage. All living organisms were destroyed. Hopf[52] observed that water containing fine suspended matter could be disinfected with small amounts of ozone where large amounts of chlorine would be required.

Scott and Lesher[51] postulated a mode of action of ozone
on *E. coli*, based on experimental data. The primary attack
of ozone was thought to be on the cell wall or membrane of
bacteria, probably by reaction with the double bonds of
lipids, and that cell lysis depended on the extent of that
reaction. Bringman[39] reported that the mode of action of
ozone differed from that of chlorine. He concluded that
chlorine selectively destroyed certain enzymes, whereas
ozone acted as a general protoplasmic oxidant. Christensen
and Giese[54] suggested that the primary locus of activity
of ozone was the bacterial cell surface. Barron[55] in 1954
hypothesized that the primary bactericidal action of ozone
was the oxidation of sulfhydril groups on enzymes. Murray
and co-workers[56] at the University of Western Ontario,
recognizing that the outermost layer of gram-negative
organisms is a lipoprotein followed by a lipopolysaccharide
layer, surmised that these layers would be first subject
to attack by ozone. They concluded that the attack by
ozone on the cell wall results in a change in cell permeability eventually leading to lysis. Smith[5] stated that
under experimental conditions where there were fewer than
1.0% survivors of *E. coli* and *Streptococcus fecalis* after
60 sec exposure to 0.8 mg/liter ozone, the unsaturated
fatty acids (mainly C_{16} and C_{18} monoenoic acids) of the
cell lipids were oxidized in the same time interval. The
lipid present in bacteria is largely confined to the cyto-
plasmic membrane. Thus, the mechanism of disinfection by
ozone is still open to question.

In 1963, ozonation was expected to become the principal
treatment for water from surface sources in Russia.[57]
Bacteriological analyses and organoleptic examinations
indicated that the water was satisfactory and ozonation
was more economical than other disinfectants.

A study by Piskunov and Sokolava[58] on the Gorki water
system produced the following conclusions: ozone doses
of 1.1 to 3.6 gm/cu.m. reduced bacterial contamination
94%, decreased color 2-1/2 times, and BOD approximately
30%. Ozone dosages of 0.27 to 1.0 gm/cu.m. following
chlorination, coagulation, and filtration effected better
results than secondary chlorination, particularly on BOD
and color removal. Laboratory tests detected an ozone
residual 2 hours after a dose of 0.55 to 0.7 mg/liter had
been applied.

Suchkov[59] found ozone to be an effective disinfectant
for drinking water containing enteropathogenic bacteria
and viruses. An organism demand effect was noted, since
more germicidal ozone was needed for higher bacterial

numbers. The same observation was made when organic matter was present in the treated water. Suchkov concluded that a reactivation phenomenon occurred several hours after ozonation if the water was incompletely disinfected (99.95% removal) but did not occur when the water was sterilized. An ozone dose of 0.2 mg/liter for 15 minutes resulted in 99.7 to 99.9% inactivation of enteroviruses.

In 1963, the use of ozone in the treatment of Loch Turret water was thoroughly investigated by Campbell,[60] and Campbell and Pescod.[61] The criterion in operating the treatment plant was to ensure that there was at least 0.1 mg/liter residual ozone in the treated water as it left the contact well, this being their index of complete disinfection. The ozone dose required to produce the 0.1 mg/liter residual was approximately 1.3 to 1.4 mg/liter with a 5-minute retention time.

In 1965, O'Donovan[62] presented a comprehensive review on ozonation. A residual ozone content of 0.1 to 0.2 mg/liter was considered sufficient to maintain a sterile effluent. The germicidal action of ozone was little affected by temperature or pH changes. Color removal was found to be incomplete. The ozone residual dissipated in a very short time; it was suggested that sterile water does not need an ozone residual such as that obtained with chlorinated waters because the disinfectant residual would not be enough to overcome accidental pollution. Bean,[63] however, flatly disagreed with O'Donovan and stated: "An experienced operator would certainly not depend upon the action of residuals of 0.1 - 0.2 mg/l to produce sterilization. Coliform organisms are seldom found where such residuals of free chlorine are maintained. Safeguards for sterility should not be abandoned; rather, they should be strengthened."

Gabovich,[64] in Russia, determined the effect of physical-chemical properties of water on the germicidal efficiency of ozonation. The attainment of a desired bactericidal effect with an increase in water temperature from 4-6° C to 18-21° C required a 60% increase in gross ozone consumption and 20% increase in net ozone consumption. Ozone gross and net consumption increased threefold with an increase in water turbidity from 4 to 50 mg/liter. The efficiency of ozone utilization decreased with increase in pH, but its bactericidal action did not diminish.

At the Eastern Sewage Works in the London Borough of Redbridge, Boucher and his associates conducted experiments on the microstraining and ozonation of wastewater efflu-ent.[65,66] Using an ozone dose of 10 to 20 mg/liter almost

all organisms were killed, although a sterile effluent
was never obtained. Chlorine followed by ozone produced
better results. In his conclusions, Boucher commented:
"Chlorination as an additional treatment to ozonation has
not produced any advantage except to destroy most of the
few organisms that sometimes survive ozonation. This is
not considered a sufficient advantage in view of its many
known disadvantages for effluent treatment, namely, the
production of chloro-derivatives which may be toxic to
fish and other aquatic life or which may produce persistent tastes, difficult to remove by subsequent waterworks
treatment, and the possibility of rapid aftergrowth of
microorganisms in a receiving river and all its attendant
problems."

In a research report on ozone treatment of secondary
effluents from wastewater treatment plants,[67] it was
determined that the product water met all USPHS chemical
and bacteriological requirements for potable water.
Virtually all color, odor, and turbidity were removed.
Oxygen-consuming organic materials, measured as COD, were
reduced to acceptable levels (<15 mg/liter). Bacteriological tests revealed that no living organisms remained.

In 1969 Carazzone and Vanini[68] found that ozone concentrations of approximately 0.5 mg/liter were virucidal
against coliphage T_2 after 5 minutes of contact.

Bender recently commented on the usefulness of ozonation as the final step in water purification.[69] He claimed
that ozone is a far more efficient germicide than chlorine,
requiring only 4 minutes of contact at 0.5 mg/liter concentration for complete inactivation of poliovirus.

In a recent symposium on ozonation in sewage treatment,
Hutchison described bacteriological studies on secondary
effluent from an extended aeration pilot plant in the
Metropolitan Sewer District of Louisville, Kentucky.[70]
Using an average applied ozone dosage of 15.2 mg/liter
for an average contact time of 22 minutes, fecal coliform
reductions of greater than 99% were achieved, resulting
in a mean fecal coliform concentration of 103 cells/100
ml, a mean total coliform concentration of 500 cells/
100 ml, and a mean fecal streptococci concentration of
8 cells/100 ml in the final effluent.

Tittlebaum and his collaborators recently studied ozone
disinfection of viruses in the Fort Southworth Pilot Plant
of the Metropolitan Sewer District in Louisville.[71] Using
F_2 bacteriophage as the model virus, they demonstrated
virtually 100% inactivation efficiency in the secondary
effluent after a contact time of 5 minutes at a total

ozone dosage of approximately 15 mg/liter and a residual
of 0.015 mg/liter. Of particular interest was the observation that the rate of inactivation was greater for F_2
bacteriophages than bacteria.

CONCLUDING COMMENTS

From the foregoing mass of literature, it is evident
that there exists much controversy, contradiction, confu-
sion, and nonfactual subjective judgment concerning the
use of ozone for disinfection of water and wastewater.
For example, it is often quite difficult to discern whether
the authors are discussing ozone residual or applied ozone
when reporting data on concentrations necessary for a
certain microbiocidal effect. In many instances, the
method of ozone analysis is sibject to critical examination. The possibility of aftergrowth of microorganisms
in a receiving water was mentioned[66] as an inherent problem
with chlorination practices. Since aftergrowth phenomena
are presumed by many to be caused by incomplete disinfection due to protective effects (such as clumps or other
factors rather than propagation of chlorine-resistant
organisms), then it follows that this problem would still
be present with ozonation provided the same protective
factors are working.
Clearly, ozone is an extremely powerful oxidizing agent,
at least as effective in all aspects as chlorine. However,
there is a great need for objective, controlled, and reproducible data on ozone disinfection technology. Numerous
questions need to be answered. What is most reliable and
precise method of ozone analysis? Does ozone really impart
an all-or-none germicidal effect, or is the effect a
typical exponential disinfection pattern similar to
chlorine? What is the effect of ozone on industrial
wastes? What effects do the ozonated effluents have on
fish and aquatic life? Some comparative work has been
done on this question by Arthur and his associates at
Duluth.[1] However, more studies with a variety of different effluents are needed. Finally, what is the objective
estimate of the costs of both installation and operation
of ozonating equipment? These questions are vital to a
true understanding of ozonation, and only through thorough
evaluation and experimentation, both on a bench-scale and
pilot-plant basis, will they be answered adequately.

REFERENCES

1. Arthur, J., "Toxic Responses of Aquatic Life to an Ozonated, Chlorinated, and Dechlorinated Municipal Effluent." Presented at the Institute on Ozonation in Sewage Treatment, Univ. of Wisconsin, Milwaukee, Nov. 9-10, 1971.
2. Kinman, R. N., "The Use of Ozone in Water Disinfection." Presented at the Sanitary Engineering Institute, Univ. of Wisconsin, Milwaukee, Mar. 9, 1971.
3. Diaper, E. W. J., "Microstraining and Ozonation of Sewage Effluents." Presented at the 41st Annual Conference of the Water Pollution Control Federation, Chicago, Ill., Sept. 1968.
4. Manley, T. C., and Niegowski, S. J., "Ozone." *Encyclopedia of Chemical Technology*, vol. 14, p. 410. John Wiley & Sons, Inc., New York (2d ed., 1967).
5. Smith, D. K., "Disinfection and Sterilization of Polluted Water with Ozone." Report AM-6704, Ontario Research Foundation (1969).
6. Kelner, A., et al., "Symposium on Radiation Effects on Cells and Bacteria." *Bacteriol. Rev.*, *19*, 22 (1955).
7. Fox, C. B., "Ozone and Antozone." J. and A. Churchill, London, 1873.
8. "The Production and Utilization of Ozone with Especial Reference to Water Purification," *Eng. News*, *63*, 488 (1910).
9. Spitta, "History of the Chlorination of Drinking Water." *Reichsgesundheitsblatt*, *3* 533 (1928).
10. Blanke, J. H., "Ozonizing Water—A French Practice." *Water Works Eng.*, *81*, 1105 (1928); *81*, 1125 (1928).
11. Baylis, J. R., "Elimination of Taste and Odor in Water." McGraw-Hill Book Company, Inc., New York and London (1935).
12. Van Ermengen, E., "Sterilization of Waters by Ozone." *Ann. Inst. Pasteur*, *9*, 673 (1895).
13. Calmette, A., and Roux, "The Industrial Sterilization of Potable Waters." *Ann. Inst. Pasteur*, *13*, 344 (1899).
14. Bulletin Municipal Official, Paris, France, p. 4722 (Dec. 6, 1932).
15. de Lipkowski, H., "Electrical Purification of Drinking Water." *Tech. Sanit. Munic.*, *32*, 54 (1937).
16. Lebout, M., "The Control of Water Purification at the City of Nice." *Tech. Sanit. Munic.*, *45*, 86 (1950).
17. Ferkinhoff, T. O., "Ozone Solves Color, Odor, and Taste Problem in Hobart Plant." *Amer. City.*, *50*, 47 (1935); *Summ. of Curr. Lit.*, *9*, 149 (1936).

18. Cox, C. R., "Significant Experiences in the Treatment of Water in New York State." J. New Engl. Wat. Wks. Assoc., *53* 444 (1939).
19. Bartuska, J. F., "Ozonation at Whiting, Indiana." J. Am. Wat. Wks. Assoc., *33,* 2035 (1941).
20. Consoer, A. W., "Use of Ozone for Water Purification." Civil Eng., *11,* 12, 701 (1941).
21. Bean, E. L., "Ozone Production and Costs," Adv. in Chem. Ser. 21, *430* (1959).
22. Smith, W. W., and Bodkin, R. E., "The Influence of Hydrogen Ion Concentration on the Bactericidal Action of Ozone and Chlorine." J. Bacteriol., *47,* 445 (1944).
23. Yanshina, M. S., "Ozonation of Water." Gigiena i Sanit., *11,* 5, 4 (1946).
24. Howlett, E., "The Ozone Method of Water Purification." Wtr. and Wtr. Eng. (G.B.), *50,* 25 (1947); J. Amer. Wat. Wks. Assoc., *39,* 1047 (Oct. 1947).
25. Whitson, M. T. B., "The Use of Ozone in the Purification of Water." J. Roy. Sanit. Inst., *68,* 448 (1948); Wat. Poll. Abs., *22,* 99 (1949).
26. Leiguarda, R. H., et al., "Bactericidal Action of Ozone." An. Asoc. Quim. Argent., *37,* 165 (1949); Wat. Poll. Abs., *22,* 268 (1949).
27. Ingram, M., and Haines, R. B., "Inhibition of Bacterial Growth by Pure Ozone in the Presence of Nutrients." J. Hyg., *47,* 146 (1949).
28. Newton, Walter L., and Jones, Myrna F., "The Effect of Ozone in Water on Cysts of *Endamoeba histolytica.*" Amer. J. Trop. Med., *29,* 5, 669 (1949); Wat. Poll. Abs., *23,* 222 (1950).
29. Novel, E., and Buffle, J. P., "Comparative Study on the Sterilizing Action of Chlorine and of Ozone on the Water of Lake Geneva Taken in the Petit-Lac." Schweiz Zeitschr. Path. in Bact., *12,* 5, 544 (1949); Wat. Poll. Abs., *22,* 218 (1949).
30. Whitson, M. T. B., "Symposium on the Sterilization of Water. (D) Other Processes with Special References to Ozone." J. Inst. Water Engrs., *4,* 600 (1950); Wat. Poll. Abs., *24,* 52 (1951).
31. Buffle, J. P., "Comparison of Bactericidal Action of Chlorine and Ozone and Their Use for Disinfection of Water." Tech. Sanit. Munic., *45,* 74 (1950); Wat. Poll. Abs., *24,* 52 (1951).
32. Bernier, H., "Ozone Sterilization of the Water Supply of Nice and Coastal Communities." Tech. Sanit. Munic., *45,* 117 (1950); Wat. Poll. Abs., *24,* 243 (1951).

33. Powell, M. P., et al., "Action of Ozone on Tastes and Odors and Coliform Organisms." J. Amer. Wat. Wks. Assoc., *44*, 1144 (1952).
34. Gubelmann, H., and Scheller, H., "Disinfection of Water by Ozone." Monatsbull. Schweiz. Ver. Gas. Wasserfachm., *33*, 53 and 99 (1953); Wat. Poll. Abs., *26* 314 (1953).
35. Rohrer, E., "Ozone and Its Application in Treatment of H_2O." Rev. Suisse Brass., *63*, 155 (1952).
36. Hettche, O., and Ehlbeck, H. W. S., "Epidemiology and Prophylaxis of Poliomyelitis with Reference to the Role of Water in its Transmission." Arch. Hyg., Berl., *137*, 440 (1953); Zbl. Bakt., I. Ref., *153*, 194 (1954); Wat. Poll. Abs., *28*, 145 (1955).
37. Dickerman, J. M., et al., "Action of Ozone on Water-Borne Bacteria." J. New Engl. Wat. Wks. Assoc., *68*, 11 (1954).
38. Holluta, J., and Unger, U., "The Destruction of *Bact. coli* Esch. by ClO_2 and O_3." Vom Wasser, *21*, 129 (1954).
39. Bringman, G., "Determination of the Lethal Activity of Chlorine and Ozone on *E. coli*." Z. Hyg. Infektionskronkh., *139*, 130 and 333 (1954); Wat. Poll. Abs., *28*, 12 (1955).
40. Wuhrmann, K., and Meyrath, J., "The Bactericidal Action of Ozone Solution." Schweiz. Z. Allgem. Pathol. Bakteriol., *18*, 1060 (1955); Wat. Poll. Abs., *29*, 223 (1956).
41. Kessel, J. F., et al., "Comparison of Chlorine and Ozone as Virucidal Agents of Poliomyelitis Virus." Proc. Soc. Exptl. Biol. Med., *53*, 71 (1943); Wat. Poll. Abs., *17*, 239 (1944).
42. Ridenour, G. M., and Ingols, R. S., "Inactivation of Poliomyelitis Virus by 'Free' Chlorine." Amer. J. Public Health, *36*, 639 (1946).
43. Lagrange, E., and Rayat, R., "One Aspect of the Battle Against Bilharzia." Societe d'Epuration et d'Entreprises, Brussels, Belgium (1952).
44. Hann, V. A., "Disinfection of Drinking Water with Ozone." J. Amer. Wat. Wks. Assoc., *48*, 1316 (1956).
45. Fetner, R. H., and Ingols, R. S., "A Comparison of the Bactericidal Activity of Ozone and Chlorine against *Esch. coli* at 1°." J. Gen. Microbiol., *15*, 381 (1956).
46. Ingols, R. S., and Fetner, R. H., "Ozone for Use in Water Treatment." Proc. Soc. Water Treatment Exam., *6*, 1, 8 (1957).
47. Stumm, W., "Some Chemical Viewpoints on the Ozonation of Water." Schweiz. Zeitschr. Hydrol., *18*, 2, 201 (1956).

48. Stumm, W., "Ozone as a Disinfectant for Water and Sewage." J. Boston Soc. Civ. Engrs., *45*, 68 (1958).
49. Miller, S., et al., "Disinfection and Sterilization of Sewage by Ozone." Advances in Chemistry Series, 21, 381 (1959).
50. Guinvarc'h, P., "Three Years Operation of the Plant Using Ozone for Disinfection of the Water Supply of Paris." L'Eau, *44*, 91 and 113 (1957); Wat. Poll. Abs., *31*, 368 (1958).
51. "Army Sterilizes Sewage by Ozone Treatment." Wastes Engr., *29*, 659 (1958).
52. Hopf, W., "Problems of Water Treatment with Ozone." Kommunalwirtschaft, *5*, 233 (1958).
53. Scott, D. B. M., and Lesher, E. C., "Effect of Ozone on Survival and Permeability of *E. coli*." J. Bacteriol., *85*, 567 (1963).
54. Christensen, E., and Giese, A. C., "Changes in Absorption Spectra of Nucleic Acids and Their Derivatives Following Exposure to Ozone and Ultraviolet Radiation." Arch. Biochem. Biophys., *51*, 208 (1954).
55. Barron, E. S., "The Role of Free Radicals of Oxygen in Reactions Produced by Ionizing Radiations." Radiation Res., *1*, 109 (1954).
56. Murray, R. G. E., et al., "Location of Mucopeptide of Selections of the Cell Wall of *E. coli* and Other Gram-Negative Bacteria." Can. J. Microbiol., *11*, 547 (1965).
57. Vakhler, B. L., "The Efficacy of Ozonation of Water from the North Donets - Donbass Canal for Drinking Purposes." Hyg. and Sanit., *28*, 3, 8 (1963); Wat. Poll. Abs., *37*, 184 (1964).
58. Piskunov, P. I., and Sokolova, N. V., "Study on Ozonation of Water at the Gorki Water System." Izv. Vysshikh Uchebn. Zavedenii, Stroit i Arkhitekt., *67*, 71 (1963); Chem. Abs., *60*, 1449f (1963).
59. Suchkov, B. P., "Studies of the Ozonation of Drinking Water Containing Pathogenic Bacteria and Viruses." Hyg. and Sanit., *6*, 24 (1964).
60. Campbell, R. M., "The Use of Ozone in the Treatment of Loch Turret Water." J. Inst. of Water Engrs., *17*, 4, 333 (1963).
61. Campbell, R. M., and Pescod, M. B., "The Ozonation of Turret and Other Scottish Waters." Water and Sewage Works, *113*, 268 (1966).
62. O'Donovan, D. C., "Treatment with Ozone." J. Amer. Wat. Wks. Assoc., *57*, 1167 (1965).

63. Bean, E. L., J. Amer. Wat. Wks. Assoc., *57*, 9, 1193, (1965).
64. Gabovich, R. D., "Experimental Studies to Determine a Hygienic Standard for Ozonation of Drinking Water." Chem. Abs., *65*, 5219H (1966).
65. Boucher, P. L., "Microstraining and Ozonation of Water and Wastewater." Proceedings of the 22nd Indus. Waste Conf., Purdue University Engr. Ext. Series 129, 771 (1967).
66. Boucher, P. L., et al., "Use of Ozone in the Reclamation of Water from Sewage Effluent." J. Inst. Pub. Health Engrs., *67*, 75 (1968).
67. Huibers, D. Th. A., et al., "Ozone Treatment of Secondary Effluents from Wastewater Treatment Plants." Report No. TWRC-4, Adv. Waste Treat. Res. Lab. and the Air Reduction Company (April, 1969).
68. Carazzone, M. M., and Vanini, G. C., "Experimental Studies on the Effect of Ozone on Viruses. I. Effect on Bacteriophage T_1." G. Batt. Virol. Immun., *62*, 11, 828 (1969).
69. Bender, R. J., "Ozonation, Next Stop to Water Purification." Power, (Aug., 1970).
70. Hutchison, R. L., "Ozonation Pilot Plant Studies at Louisville." Presented at the Institute on Ozonation in Sewage Treatment, Univ. of Wisconsin, Milwaukee, Nov. 9-10, 1971.
71. Tittlebaum, M. E., et al., "Ozone Disinfection of Viruses." Presented at the Institute on Ozonation in Sewage Treatment, Univ. of Wisconsin, Milwaukee, Nov. 9-10, 1971.

CHAPTER VI

OZONE GENERATION AND ITS RELATIONSHIP

TO THE ECONOMICAL APPLICATION OF

OZONE IN WASTEWATER TREATMENT

Harvey M. Rosen

INTRODUCTION

Ozone is a relatively unstable gas produced commercially
in large volume by the reaction of an oxygen containing feed
gas in an electric discharge called a corona. Another commercial technique is the use of ultraviolet energy rather
than corona energy, but this method produces only low volume,
low concentration ozone and so is applicable only to very
small systems.

The instability of ozone with respect to decomposition
back to oxygen dictates the need for an on-site production
facility when ozone is used. This in turn dictates the
need for a cost-efficient, space-efficient, low-maintenance
installation if ozone is to be applied in large wastewater
applications.

Currently, all large-scale commercial ozone generators
operate on the corona discharge principle. The only viable
technique which may compete with corona for the future efficient production of ozone is radiochemical, that is, the
use of nuclear power to supply the energy needed to convert
oxygen to ozone. Work performed at Brookhaven National
Laboratories[1] indicates the feasibility of producing hun-
dreds of tons of ozone per day by a chemonuclear process.
Only at these high levels of production does this method
appear to compete economically with corona discharge for

ozone synthesis. While the chemonuclear technique is efficient in theory, a good deal more work will be required to
make it a practical reality.

OZONE GENERATION THEORY

When high-voltage AC is imposed across a discharge gap
in the presence of an oxygen-containing gas, ozone is produced (Figure 33). This basic method of production is
inherently inefficient. Only about 10% of the energy
supplied is used to make ozone. The remainder is lost as
light, sound, and primarily heat. Unless heat is removed
efficiently, the ozonator gap acts as an oven, and high
temperatures build up in the discharge space and at the
dielectric surfaces. If temperatures are allowed to build
up, ozone yield will suffer, since decomposition of ozone
is very temperature sensitive, and the dielectric characteristics can be affected to the point of causing dielectric
puncture. Therefore, an efficient method of heat removal
is essential.

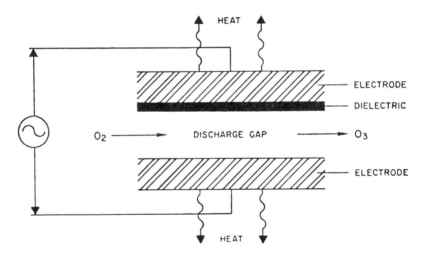

Figure 33. Basic Ozonator Configuration

When a clean, dry, oxygen-rich gas is fed to the ozone
generator and an efficient method of heat removal is available, then the production of ozone from corona under
optimum conditions can be represented by the following
relationships.[2]

$$V \alpha\ pg \qquad\qquad (1)$$

$$(Y/A)\ \alpha\ \frac{f\varepsilon V^2}{d} \qquad\qquad (2)$$

Where:
  (Y/A) is ozone yield per unit area of electrode surface
    under optimum conditions
  V is voltage across the discharge gap (peak volts)
  p is gas pressure in the discharge gap (psia)
  g is width of the discharge gap
  f is frequency of the applied voltage
  ε is dielectric constant of the dielectric
  d is thickness of the dielectric

Therefore, to optimize ozone yield (taking into consideration practical material and engineering requirements),
the following conditions should exist:

1. The pressure/gap combination should be constructed
so that the voltage can be kept relatively low while reasonable operating pressures are maintained. Keeping
voltage low protects the dielectric and/or the electrode
surfaces from high voltage failure. Operating pressures
from 10 to 15 psig are useful in treatment of wastewater
if ozone is to be injected at the bottom of standard waste
water treatment tanks. Some energy is necessary to bring
the water and gas into contact in any case, and if ozone
is produced under pressure this energy can be used in the
contact step.

2. A thin dielectric material with a high dielectric
constant should be used. Equation (2) indicates that
both these properties of the dielectric allow for high
yield efficiency. Some form of glass is the only practical
dielectric material from the point of view of high dielectric constant, availability, and cost. It is necessary to
maintain a high dielectric strength to minimize dielectric
puncture while minimizing dielectric thickness to optimize
ozone yield and facilitate heat removal from the system.

3. High frequency AC should be used. High frequency is
less damaging to the dielectric surfaces than high voltage.
This decreases maintenance requirements and increases the
useful life of the machine, while producing increased ozone
yields. There is obviously a trade-off between voltage and
frequency in the energy source. Voltage is appealing because it appears in the equation to the second power.
However, this advantage of voltage can be deceiving when
considering the practical aspects of dielectric failure

at high voltages. An ideal balance of theoretical and practical considerations is to maintain a dielectric thickness as small as possible (while minimizing practical problems in using thin material) to construct a practical ozone generator, and in preventing dielectric puncture at reasonable operating voltages.

4. Heat removal should be as efficient as possible. Solely relying on the gas flow through the gap to remove heat allows production of ozone in concentrations of less than a few tenths of a percent. If ozone is required in higher concentrations, an additional means of cooling is necessary.

Therefore, an essential point to be considered in the optimization process is the removal of as much heat as possible from the system. The voltage or frequency can be continually increased to produce ozone, but the process is efficient only as long as it is practical to remove the excess heat produced by the more demanding conditions.

COMMERCIAL OZONE GENERATORS

With this background, commercially available corona-type generators can be examined. There are three basic types: the Otto plate, the tube, and the Lowther plate. While some variations, such as vertical or horizontal mounting of ozonator cells, are available these types represent the basic configurations.

The Otto Plate Type

The original design for this plate generator was developed by Otto in the beginning of the twentieth century, and this basic unit is still being used, primarily in potable water treatment applications. The ozonator is made up of a number of sections arranged in the following sequence: a cast-aluminum, water-cooled block which acts as the ground electrode; a glass plate dielectric; an air space; another glass dielectric; and a high voltage stainless steel electrode. A complete "unit" would include the mirror image of dielectrics, air space, and a grounded, water-cooled electrode (Figure 34).

Air is blown into the ozonator and enters the discharge gap where conversion to ozone takes place. The ozonized air is drawn through a manifold pipe formed by holes cut in the center of each of the electrodes and dielectrics.

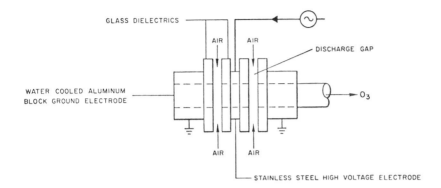

Figure 34. Otto Plate-Type Generator Unit

This type of gas distribution system is one drawback of the Otto plate-type generator, since operation is limited to low pressure.

The Tube Type

The tube-type generator (Figure 35) is composed of a number of tubular units. The outer electrodes are stainless steel tubes fastened into stainless steel tube spacers and surrounded by cooling water. Centered inside the stainless steel tubes are tubular glass dielectrics whose inner surfaces are coated with a conductor which acts as the second electrode.

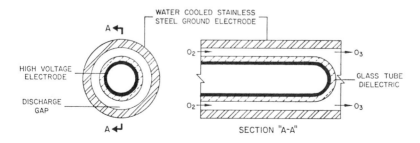

Figure 35. Tube-Type Generator Unit

The stainless steel outer tubes are arranged in parallel
and are sealed into a cooling water distribution system.
The group of water-cooled tubular units are then enclosed
in a gas tight "iron lung" so that air or oxygen can be
fed to the ozonator at one end of the tubes and ozone collected at the other. The glass tube is sealed so that
the feed gas passes only through the discharge gap.

The Lowther Plate Type

This generator, although a plate type, is significantly
different from the Otto plate type. The Lowther generator
is air cooled, and operates on either air or oxygen feed.
The basic unit is a gas-tight "sandwich" made up of an
aluminum heat dissipator, a steel electrode coated with a
ceramic dielectric, a glass spacer to set the discharge
gap, and a second ceramic coated steel electrode with an
oxygen inlet and an ozone outlet which exit through a
second aluminum heat dissipator (Figure 36). These components are pressed together in a frame and manifolded for
oxygen and ozone flow. Cooling is accomplished by a fan
moving ambient air across the heat dissipators.

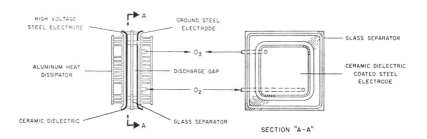

Figure 36. Lowther Plate Generator Unit

QUALITY OF FEED GAS

There are two more aspects of corona generation which
must be discussed before a final comparison can be made.
The first is the composition of the oxygen-containing
feed gas. If it is assumed for now that only nitrogen
will be present with the oxygen in the feed gas, 2 to
2-1/2 times as much ozone will be produced from a stream
of 100% oxygen as from an air stream (assuming all other

conditions are equal). (See Figure 37.) For reasons of
economics, to be discussed in more detail later, it is
essential in any large application to feed oxygen or
oxygen-enriched air to the ozonator.

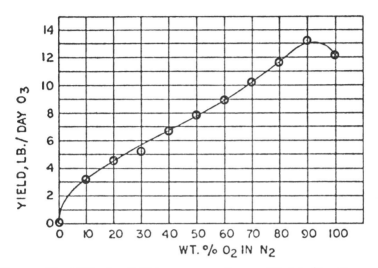

Figure 37. Effect of Nitrogen on Ozone Yield (Lowther
         Generator)

Whether feeding oxygen, oxygen-enriched air, or air to
the ozonator, the gas stream must be dry since ozone
decomposition is accelerated by the presence of moisture.
In addition, nitrogen present in the gas stream combines
with oxygen, ozone, and water in the corona to produce
nitric acid, which can further reduce ozone yield and
damage the material parts of the ozonator.

All corona generation equipment is made up of individual
cells which are manifolded together to produce ozone in
quantities. These cells may be vertical or horizontal but
the configuration seems to make little difference in the
overall efficiency of a particular type of cell. Based
on the requirements for efficient ozone generation by
corona, typical operating characteristics of the commercial type generators can be compared. (See Table 9.)

Referring back to Equations (1) and (2) and comparing
the data in Table 9, it becomes evident that the Lowther
generator is designed to produce ozone by corona in the
most efficient and practical manner. The problem boils
down to one of efficient heat removal. In the Otto plate-
type and tube-type systems, all heat must be removed from

Table 9. Comparison of Commercial Ozonators. Typical
Ozonator Operating Characteristics

	Otto	Tube	Lowther
Feed	air	air, oxygen	air, oxygen
Dew point of feed, °F.	-60	-60	-40
Cooling	water	water	air
Pressure, psig	0	3-15	1-12
Discharge gap, in.	0.125	0.10	0.05
Voltage, kv. peak	7.5-20	15-19	8-10
Frequency, Hz	50-500	60	2000
Dielectric thickness, in.	0.12-0.19	0.10	0.02

one electrode surface. This makes these modules severely
heat limited, since heat from the uncooled electrode
surface must pass across the relatively large discharge
gap and across relatively thick dielectrics before it can
be removed from the system. Another result of high yield
efficiency is high space efficiency so that savings in
real estate are also available.

These characteristics also result in lower power requirements than the generators based on older technology
(Table 10).

Table 10. Ozone Generator Power Requirements (for 1% ozone)

Type	Air Feed kwh/lb.	Oxygen Feed kwh/lb.
Otto	10.2	-
Tube	7.5-10	3.75-5.0
Lowther	6.3- 8.8	2.5 -3.5

ADVANTAGES OF OZONE IN WASTEWATER TREATMENT

A great deal of research and development has gone into producing less costly, more efficient ozone generators because ozone has such unique properties which offer advantages in wastewater treatment. Among them are:

1. Ozone is a powerful oxidizing agent which has twice the chemical oxidation potential of chlorine in its reactive form as hypochlorite ion. As a result, more complete oxidation can be expected from ozonation than from chlorination.

2. Many ozone reactions are very rapid. Unlike chlorine, ozone does not always have to go into solution before reacting. In the case of disinfection, there is some evidence that the lysing reaction occurs between gaseous ozone and the microorganism. Faster reaction times can mean shorter contact times to reach required effluent contaminant levels and, as a result, reduce capital for contactors.

3. Ozone is a highly efficient germicide. This results in surer bactericidal and viricidal action with shorter contact times and less sensitivity to pH and temperature than for chlorine. The reactions with viruses are so rapid that they are difficult to study analytically. However, recent work at Louisville[3] has shown a more rapid and sure kill of viruses than of bacteria by ozone. In both cases, ozonation was an improvement over chlorination in terms of rate and sureness of disinfection.

4. It leaves a beneficial oxygen residual as a reaction product. Ozonation for COD reduction, as demonstrated in the Blue Plains pilot plant[4] with oxygen feed and recycle to the ozone generator, showed a DO of 40 ppm in the effluent. The use of air as the ozonator feed would, of course, leave a lower residual but would still increase the DO.

5. As a class, oxidized or partially oxidized products are generally less toxic than chlorinated or unreacted species. This is a generality which requires further testing. However, comparison of the toxicity of ozonated versus chlorinated effluents[5] tends to support the generality.

As an oxidizing agent, less reaction time is usually required than for other chemical oxidizers in a particular application because of its higher reactivity. However, this efficiency is partially offset by the decomposition of ozone in aqueous solutions in which it has a half-life

of approximately twenty minutes. Since many oxidation
reactions and disinfection reactions are rapid, efficient
use of ozone must be accomplished with short contact times
and little loss due to decomposition.

A major advantage results from the use of ozone as compared with other oxidizing agents or disinfectants: a
wastewater with a low toxicity level and a high DO concentration is produced by ozone treatment. Chlorination may
produce highly toxic chlorinated organics and chloramines
(chlorinated organics as a class are the most toxic group
of organics found in wastewater) while ozone produces
fully or partially oxidized organics and oxygen.

OZONE IN WASTEWATER TREATMENT

Ozone can be used in several places in a wastewater
treatment plant. These may include:

1. Disinfection of treatment plant effluents
2. Tertiary treatment
   a. Reduction of COD and removal of BOD
   b. Disinfection
   c. Increased DO
   d. Reduction of color and odor
   e. Decrease of turbidity
3. Sludge treatment
   a. Oxidation of secondary sludge for partial or
      complete volatilization of organics.
   b. Partial oxidation and lysing to make bacteria
      and other organics available as food in recycle
      to activated sludge.
   c. Breaking up filamentous bacterial growth and
      colloid structure to allow easier dewatering.
4. Combined treatment with activated carbon, filtration, ultrasonics, or other chemicals.
5. Odor control

Disinfection

Each of these is worth taking in turn. In the United
States, the disinfection of treatment plant effluents will
probably be the area where ozone will have its most immediate impact on a large scale. California has recently
completed a study in which it was shown that chloramines
are toxic to aquatic organisms.[6] Wyoming, Michigan, in
conjunction with the Environmental Protection Agency, is

conducting a study to determine the toxicity to fish of
disinfected secondary effluents. Streams which have been
(a) chlorinated, (b) chlorinated then dechlorinated with
sulfur dioxide, and (c) ozonated will be compared. The
Wyoming study is a large-scale research project designed
to confirm earlier conclusions on the toxicity of chlorinated
effluents at Wyoming[7] while trying to establish alternatives
to chlorine disinfection. A similar study is substantially
underway at the EPA regional laboratory in Duluth,
Minnesota.[5] The Michigan study will include experiments
on the effective contact of ozone with secondary effluents
to economically accomplish disinfection without losing
ozone to ozone-demanding wastes in the effluent stream.

Ozone could be used on a purely economic basis for
wastewater disinfection. New York City pays 23¢ per
pound of available chlorine by purchasing calcium hypochlorite for disinfection. Chicago and other major cities
are in the same position as a result of legal restrictions
against the storage of tank car quantities of liquid
chlorine in populated areas. With economics as the driving
force, Chicago has completed a study on ozone disinfection[8]
and will probably extend the study to include organics reduction in a larger pilot plant. A decision on this is
still pending.

Ozone is such a powerful oxidizing agent that it is very
difficult to control its reactions in a high TOC containing
wastewater. That is, when it is supplied in an adequate
concentration for a sufficient contact time, it will react
with almost anything. As a result, selectivity is hard to
achieve except through kinetic control. Studies are cu-
rently underway to elucidate the mechanism of disinfection
by ozone as compared with the other possible reactions
with which disinfection is in competition. Hopefully, the
mechanisms or rates of reaction are sufficiently different
to allow selective application. Qualitatively, this seems
to be the case since historically a major problem in
studying ozone disinfection has been that reactions occur
so rapidly that analytical techniques have limited the
investigations.

Tertiary Treatment

An EPA pilot study has just been completed by Air Reduction Company on the ozonation of secondary effluents
for "complete" tertiary treatment.[4] In this context,
"complete" means that final polishing produces an almost
potable water with the primary exception of the removal

of some dissolved and suspended solids. Reduction of
BOD, COD, disinfection, and aeration all occurred in the
ozonated effluents. While final data from the pilot study
are not yet published, it is evident from the preliminary
results that significant COD reductions to below 15 ppm
can be achieved at initial ozone dosages of 0.7 to 3.0
ppm ozone per ppm COD removed.

It was determined from a previous feasibility study[4],[9]
that the estimated operating cost for treatment in a 10
MGD plant is 7.7¢/1000 gal. compared with 8.3¢/1000 gal.
for carbon adsorption. This cost assumes an ozone utilization efficiency of 80%, and does not include the added
savings resulting from simultaneous disinfection or the
value in the added benefit of a high DO as a residual of
ozone treatment.

The Airco pilot scale investigation at Blue Plains was
a research study and the pilot plant was designed for
flexibility, not economy. Studies in the pilot plant
concerned ozonation of regular activated sludge effluent,
UNOX system effluent, and physical-chemical treated
effluents. Ozonation dosages were controlled by addition
in up to six stages with gas dispersion injectors. In
general, contact time in each stage was about ten minutes
for a total treatment time of one hour. Dosage rates were
varied to achieve COD levels of 15 mg/liter. In most of
the effluents studied, this meant that BOD was reduced to
zero mg/liter, complete disinfection was accomplished,
and a high DO resulted.

The Airco study is only the beginning of many which
must be conducted with ozone in order to establish design
criteria and to minimize treatment costs before large
scale acceptance is forthcoming. New ozone generation
technology, and optimizing the application of ozone to
get 90% utilization efficiency, could reduce the costs of
"tertiary treatment" to less than 7.0¢/1000 gal.

Sludge Treatment

Another application for ozone, about which less is
known, is sludge treatment. Partial oxidation to remove
volatiles makes the sludge more desirable as land fill
and allows its application on the land to depths of several
feet rather than several inches for untreated sludge. In
some areas sludge handling and removal is a major cost
factor in wastewater treatment. In Denver, sludge removal
costs the city $50 to $60 a ton, and Denver Metropolitan
Sanitary District and Martin-Marietta Corporation are

investigating the economics of complete oxidation of organic sludge to volatiles using oxygen/ozone mixtures. Early results indicate that small amounts of ozone appear to catalyze the oxidation process. If this were the case, and stoichiometric amounts of ozone were not required, the use of oxygen containing ozone for sludge oxidation could become economical.

A second potential use in activated sludge is the partial oxidation of organics and lysing of bacteria. The desired result of this process is to make the degraded material available as food through a recycle to the activated sludge process, thereby achieving a net reduction in solids for disposal. The concept of treating bacteria to make them available as food for other bacteria is not new, but the use of ozone to accomplish this is just being investigated.

A third application in sludge treatment is the attack of the filamentous bacterial growth which makes sludge difficult to dewater. Small dosages of ozone applied selectively could destroy the colloidal structure that holds water so tightly. Based on this possibility, work is to be performed by the E.P.A. in its installation at Blue Plains, which is aimed at conditioning of sludge with ozone to aid dewatering.

Ozone Combined with Other Unit Processes

In some instances, where COD consists of difficult to oxidize organics, ozone in combination with activated carbon to polish and disinfect and deodorize the column is a possibility. Carbon columns increase in capacity when some biological growth occurs, but this can cause production of hydrogen sulfide. Ozone can be used to control the growth and eliminate odors due to H_2S. In those cases, where the BOD and COD of a secondary effluent are very high, it would probably be economical to consider a chemical preclarification step before ozonation. In other cases, ozone in combination with a mechanical filtration step such as microstraining will be applied.[8] Recently work with ozone in combination with ultrasonics has been reported in which phosphorus and nitrogen appear to be removed from solution by a yet unknown mechanism.[10] Other work combines ozone with ultrasonics and catalysts for phenol and other organic oxidations.[11] Another study combines ozone with froth flotation for disinfection and clarification.[12]

Odor Control

Lastly, ozone for odor control is an established process, and cities such as Cedar Rapids, Iowa, have been
using it successfully for a number of years in the municipal wastewater treatment plant. This application is
becoming more popular since light plastic covers for
outside treatment tanks have become available.

OZONATION SYSTEMS

In thinking of wastewater treatment as opposed to
sludge treatment or odor control, consideration must be
given to the system requirements for this application
which go far beyond an efficient ozone generator. Four
systems can be generalized for the application of ozone
in wastewater treatment. These are:

IA—Once through air
IB—Once through oxygen-enriched air
IIA—Recycled pure oxygen
IIB—Recycled oxygen enriched air

Open-Loop Ozonation

In system IA (Figure 38), compressed air is cooled to
remove moisture and is fed to a dryer. The dry air is fed
to the ozonator and the ozone-air solution (1-3 weight %
ozone) is mixed with the wastewater in a contactor. The
treated water and gases leave the contactor separately.
Any excess ozone in the water soon decomposes to oxygen
while the ozone in the waste gas is destroyed by heat or
by chemical or catalytic decomposition. This system is
applicable to only small installations such as shipboard
systems because of the basic inefficiency in using air
instead of oxygen to feed the generator. This type of
system is also used for odor control where a large volume
of gas is to be treated and the feed gas becomes too
dilute to recover and recycle.
System IB (Figure 38) is similar to system IA but is
somewhat more cost-efficient because air has been replaced
by an oxygen-enriched feed stream. The use of a pressure
swing separation technique for producing oxygen-enriched
air will reduce the capital and operating costs of the
ozonator. In an economic trade-off, the cost of producing
oxygen by pressure swing separation is less than the cost

(I) <u>AIR SYSTEM</u>

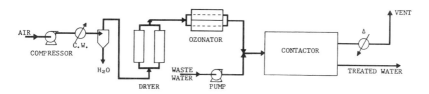

(II) <u>OXYGEN-RICH AIR SYSTEM</u>

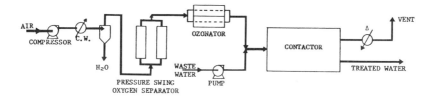

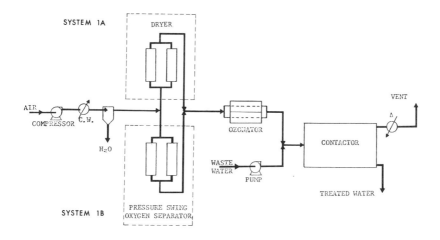

Figure 38. Open-Loop Ozone Treatment Systems

of producing ozone. However, since the nominal concentration of ozone produced is 2% by weight in the feed stream,
50 pounds of feed gas are required to produce a pound of
ozone product. To arrive at an economic balance, the cost
of oxygen-enriched air generation at a particular concentration of oxygen must be compared with the cost of the
ozone yield available from that concentration. Pressure
swing air enrichment and its advantages for ozone generation will be discussed in the next section.

Closed-Loop Ozonation

The third system, IIA (Figure 39), for ozone treatment
is similar to system IA, but, instead of air, oxygen is
used as the feed gas to the system, and offgas is recycled
to the front end of the loop. This arrangement necessi-
tates an additional deaeration step which removes nitrogen
from the wastewater before treatment so it does not build
up in the gas recycle. This is basically the flow diagram
for the Airco pilot plant.

The ultimate system is shown in flow sheet IV (Figure
39). In this case, air is enriched to perhaps 40% oxygen
in the start-up cycle. In each successive cycle, the off-
gas is recycled through the pressure swing separator; is
cleaned, dried, and enriched in oxygen; and any excess
ozone is decomposed by the catalytic effect of the molecular
sieves used in the separator. The system can be programmed
so that the economic point of ozone generation versus oxygen
generation is achieved. That point is likely about 80%
oxygen, the other 20% consisting of nitrogen, carbon
dioxide, and argon.

By mixing adsorbents in the pressure swing separator,
nitrogen, carbon dioxide, and water will be removed on
each cycle, and the oxygen increased. Air, of course, is
added for makeup. At a combined level of 10% or less,
carbon dioxide and argon have little effect on ozone production efficiency, and small quantities of nitrogen
appear to have a positive effect. A closed-loop system
using the pressure swing separator provides:

1. oxygen enrichment;
2. removal of nitrogen and carbon dioxide;
3. drying to -100° F dew point; and
4. catalytic decomposition of excess ozone.

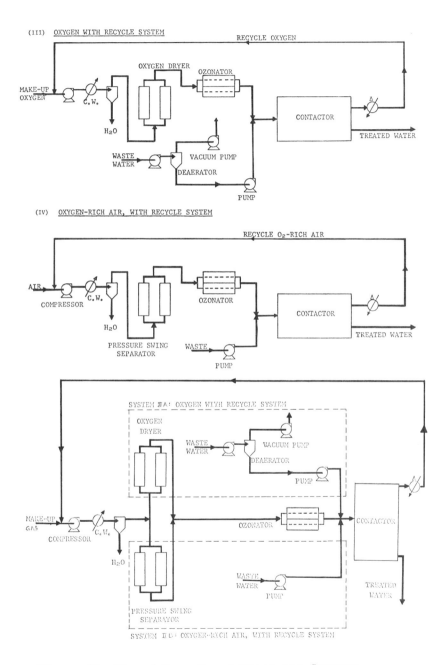

Figure 39. Closed-Loop Ozone Treatment Systems

Pressure Swing Separation

Pressure swing separation for oxygen enrichment operates this way. In the first half-cycle (Figure 40), compressed air is introduced into a tower filled with a proprietary molecular sieve adsorbent designed to be selective for nitrogen adsorption. The surface characteristics of this sieve cause nitrogen molecules to be preferentially trapped within the solid adsorbent particles, and oxygen passes through the tower to be collected as product.

At the same time, a second column, which was saturated with nitrogen under high pressure in the preceding step, is depressurized and nitrogen is released under the reduced pressure. Some of the oxygen product is passed through this second tower to accelerate the removal of nitrogen by purging.

In the second half-cycle, compressed air is switched to the now-depressurized column while the column initially under pressure is being depressurized. Now nitrogen is being trapped in the second column which is producing oxygen-rich product while the first tower is being regenerated. At the end of the second half-cycle, the second tower is saturated with nitrogen and the first tower has completed its release of nitrogen. The towers are now ready for the next half-cycle, which is the same as the first half-cycle already described.

The addition of a surge tank, which collects and mixes the nitrogen-deficient outputs from both towers, produces a continuous stream of oxygen-enriched air. The product gas will be of constant composition, containing from 40 to 95% oxygen. Variation in adsorbent, bed size, and configuration, pressure, and cycle times provide the means for producing the required gas volume at the desired oxygen concentration. The concentration range from 85 to 95% (Figure 31), however, is costly to attain (which is not unexpected when an incremental increase approaching unity is attempted). For ozone generation, the use of 80% oxygen results in approximately the same generation efficiency as 100% oxygen.

A further benefit of using the pressure swing oxygen generator to feed the ozonator is that it produces a minus 100° F dew point product. While the molecular sieve has been optimized for nitrogen-oxygen separation, it is still more selective for water (sieves are widely used in drying applications). While little efficiency is gained by going to such a low dew point, it was discussed previously that the drier the feed stream to the ozone

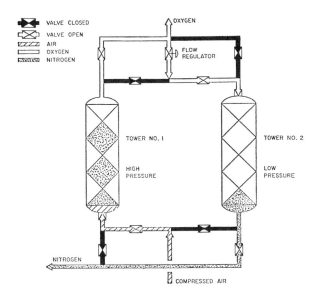

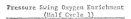

Pressure Swing Oxygen Enrichment
(Half Cycle 1)

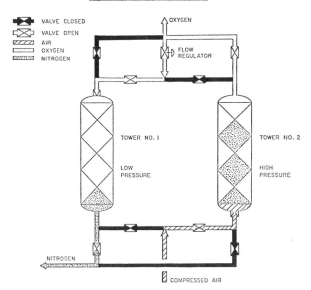

Pressure Swing Oxygen Enrichment
(Half Cycle 2)

Figure 40. Pressure Swing Oxygen Enrichment

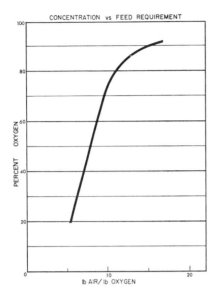

CONCENTRATION vs FEED REQUIREMENT

Figure 41. Oxygen Enrichment
by Pressure Swing
Separation

generator, the better from the standpoint of equipment
maintenance.

CONTACTING SYSTEMS

   After a decision on the use of one of the open-loop or
closed-loop systems has been made, the contactor design
must be considered to maximize the ozone transfer efficiency and to minimize the net cost for treatment. Just
as every wastewater is different so must the contactors
be different, and the importance of the design cannot be
stressed enough. However, there has been little work
done in this area for gas/liquid contacting in wastewater
treatment applications. Much of the work done for fluids
of commercial value is not applicable to waste treatment
technology because of economic limitations. When considering milk, wine, beer, etc., a process step can cost
cents per gallon, but for wastewater the process cost
must be less by a factor of a thousand, that is cents per
1000 gallons. There is a great deal of room in this area
for chemical engineers to apply their know-how to designing specific contact systems for the application of ozone
to wastewater.
   To define the contactor the following must be specified:

1. The job to be done: disinfection, BOD or COD reduction to a particular level, etc.
2. Relative rates of competitive reactions: chemical oxidation, lysing bacteria, decomposition of ozone in aqueous solutions, etc.
3. Mass transfer rates for ozone dissolution in wastewater.
4. Characteristics of the wastewater: total suspended solids, organic loading, pH, temperature, etc.
5. Total operating pressure of the system.
6. Concentration of ozone employed: 1 to 6%.

Further specific considerations on the contacting system include:

1. Contactor type
   a. packed bed
   b. sparged column
   c. sparged column with mixing, etc.
2. Number of contact stages. (Determined by extent of treatment; all stages may not be the same type.)
3. Methods of gas dispersion and mixing within stages
4. Configuration of contactors
5. Points of gas/liquid contact
6. Cocurrent or countercurrent mixing
7. Materials of construction

Thus, designing an ozonation treatment plant even when only considering "tertiary" municipal treatment requires a thorough knowledge of the individual situation to optimize a least-cost system. In many future plants, oxygen will be used for aeration in the activated sludge process, for aerobic digestion of sludges, and for incineration. How much more complex will be the system designed to minimize oxygen-ozone applications costs!

The many advantages of ozone in the treatment of many difficult industrial wastes such as phenols and cyanides have not been mentioned. The problems in the treatment of industrial wastewaters produce a whole new set of design criteria often related to water recycle. Ozone and oxygen are excellent in this respect since oxygen is the only residual left by both gases.

Advanced technology for economical oxygen and ozone generation is now available. Some systems for their application have been developed and demonstrated. However, only the surface has been scratched in the development of applications technology not only in the municipal area but also in industrial water pollution control.

REFERENCES

1. Steinberg, M., and Beller, M., "Ozone Synthesis for Water Treatment by High Energy Radiation." <u>Chem</u>. <u>Eng</u>. <u>Prog</u>. <u>Ser</u>. <u>No</u>. 104, *66*, 205 (1970).
2. Lowther, F. E., "Theoretical Considerations in the Design of an Ozonation System." To be published.
3. Tittlebaum, M. E., et al., "Ozone Disinfection of Viruses." Presented at Institute on Ozonation in Sewage Treatment, Univ. of Wisconsin, Milwaukee, November 9-10, 1971.
4. McNabney, R., and Wynne, J., "Ozone: The Coming Treatment?" <u>Water and Wastes</u> <u>Eng</u>., *8*, 46 (Aug. 1971).
5. Arthur, J. W., "Toxic Responses of Aquatic Life to an Ozonated, Chlorinated and Dechlorinated Municipal Effluent." Presented at the Institute on Ozonation in Sewage Treatment, Univ. of Wisconsin, Milwaukee, November 9-10, 1971.
6. Esvelt, L. A., Kaufman, W. J., and Selleck, R. E., "Toxicity Assessment of Treated Municipal Wastewater." 44th Annual Conference of the Water Pollution Control Federation, San Francisco, California, October 3-8, 1971.
7. Zillich, J. A., "Toxicity of Combined Chlorine Residuals to Freshwater Fish." <u>J</u>. <u>Water</u> <u>Poll</u>. <u>Contr</u>. <u>Fed</u>., *44*, 212 (1972).
8. Zeny, D., "Ozonation Pilot Plant Studies at Chicago." Presented at the Institute on Ozonation in Sewage Treatment, Univ. of Wisconsin, Milwaukee, November 9-10, 1971.
9. Huibers, D. Th. A., McNabney, R. and Halfon, A., "Ozone Treatment of Secondary Effluents from Waste-Water Treatment Plants." Report No. TWRC-4, Robert A. Taft Water Research Center.
10. ———, "Ozone and Sonics. . . ." <u>Chemical Week</u> (September 22, 1971), 73.
11. Feasibility Studies of Applications of Catalytic Oxidation in Wastewater, U.S. Environmental Protection Agency, Water Pollution Control Research Series, Report 17020 ECI 11/71.
12. Foulds, J. M., et al., "Ozone Generated Froth for Sewage Treatment," <u>Water</u> <u>and</u> <u>Sewage</u> <u>Works</u>, *118*, 80 (Mar. 1971).

CHAPTER VII

OZONE IN WATER DISINFECTION

Riley N. Kinman

HISTORY

Ozone was first noted by Van Marum in 1785[1] in the vicinity of an electrical machine. The name "ozone" is derived from the Greek *ozein*, "to smell"; and this gas, colorless at room temperature, owes its name to its characteristic odor. Schonbein reported in 1840 that the odor was due to a new substance, and several years later this substance was shown to be triatomic oxygen, O_3. Ozone's first important commercial use was in the disinfection of water. Several experimental plants were in use as early as 1892, but the first really important plant went into operation in Nice, France, in 1906. Some 100 municipal installations were in operation in France, and 30 to 40 more were reported in other countries by 1936. It is interesting to note that ozone underwent its peak development for water disinfection right after its commercial introduction.

In July 1940,[2] Whiting, Indiana, put ozone generating equipment into service, because of the tastes and odors produced when their Lake Michigan raw water was chlorinated. This city has the longest operating experience with O_3 of any American city. The Whiting system will be discussed later; however, it should be pointed out here that Whiting does not use the ozone for disinfection but as a pretreatment ahead of the addition of chlorine and ammonia.

In 1949,[3] Philadelphia began operation of the world's largest O_3 plant for the removal of tastes, odors, and manganese from the grossly polluted Schuylkill River.

However, the use of O_3 was suspended in 1959 because chlorination was found to be less expensive when the plant was expanded.

In 1950,[1] part of the St. Maur supply for Paris, France, was placed on O_3, principally because of complaints about the taste and odor of the water. Paris ozonates about one-third of its supply, and it has been reported[4,5] that approximately 136 municipal water plants serving 8 million people are currently using ozonated water in France. England and Germany[6] were also reported to have several municipal installations using O_3 as the water disinfectant. Canada and Mexico use O_3 to some extent, but very little use of O_3 has been made in U.S. water plants. Several reasons for this apparent lack of enthusiasm on the part of U.S. engineers in the use of O_3 disinfection will be discussed later in this paper.

PHYSICAL PROPERTIES

Ozone (O_3, molecular weight 48) with its characteristic pungent odor is colorless at room temperature and condenses to a dark blue liquid. It is generally encountered in dilute form in a mixture with oxygen or air. Liquid O_3 is very unstable and will readily explode. Concentrations of O_3 in air-oxygen mixtures above 30% are easily exploded. Explosions may be caused by trace catalysts, organic materials, shocks, electric sparks, or sudden changes in temperature or pressure. Ozone absorbs light in the infrared, visible, and ultraviolet at certain wavelengths. It has an absorption maximum at 2537-Å.

Ozone is more soluble in water than is oxygen, but, because of a much lower available partial pressure, it is difficult to obtain a concentration of more than a few milligrams per liter under normal conditions of temperature and pressure. Figure 42 depicts the theoretical solubility of O_3 in water. Ozone supposedly decomposes in water, but this is probably due to its strong oxidizing ability rather than simple decomposition. Neutral salts and hydroxyl ions accelerate this decomposition. Ozone is much more soluble in acetic acid, acetic anhydride, propionic acid, propionic anhydride, dichloroacetic acid, chloroform, and carbon tetrachloride than it is in water.

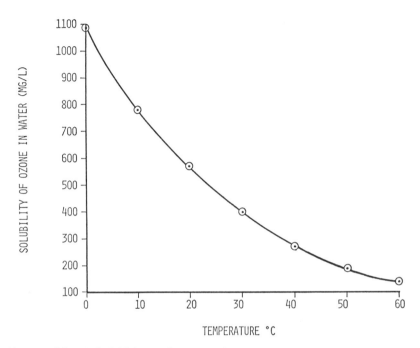

Figure 42. Solubility of Ozone in Water

CHEMICAL PROPERTIES

Ozone is reported[1] to be naturally unstable and to
decompose to ordinary oxygen slowly. Heat accelerates
decomposition, and decomposition is instantaneous at
temperatures of several hundred degrees C. Moisture,
silver, platinum, manganese dioxide, sodium hydroxide,
soda lime, bromine, chlorine, and nitrogen pentoxide
catalyze decomposition. Ozone is also decomposed photo-
chemically. From a practical standpoint, decomposition
is slow enough to permit the use of ozonized air or oxygen
streams for water disinfection.

Ozone is a powerful oxidizing gas. Only fluorine, F_2O,
and O have higher electronegative oxidation potentials.
Table 11 contains some of the more important disinfecting
species with their respective oxidation potentials referred
to the hydrogen electrode at 25° C and unit hydrogen ion

Table 11. Properties of Certain Disinfectant Species in
H_2O (Oxidation Potential Refers to the Hydrogen
Electrode at 25° C and at Unit Hydrogen-Ion
Activity)

Disinfectant Species	Weight per Mole (gr.)	Molecules mg/liter	Oxidation Potential (volts)
O_3	48.00	$1.26 \cdot 10^{19}$	-2.07
HOBr	96.91	$6.21 \cdot 10^{18}$	-1.59
HOCl	52.46	$1.15 \cdot 10^{19}$	-1.49
HOI	143.91	$4.19 \cdot 10^{18}$	-1.45
Cl_2	70.90	$8.50 \cdot 10^{18}$	-1.36
Br_2	159.81	$3.77 \cdot 10^{18}$	-1.07
NH_2Cl	51.47	$1.17 \cdot 10^{19}$	-0.75
NH_2Br	95.93	$6.27 \cdot 10^{18}$	-0.74
I_2	253.81	$2.38 \cdot 10^{18}$	-0.54

activity. Ozone retains this strong oxidizing ability in aqueous solution. It attacks most metals except gold and platinum. It oxidizes ferrous iron to ferric, manganous dioxide to manganese or permanganate, chromous ion to chromate or dichromate, arsenite to arsenate, sulfide to sulfate, and nitrite to nitrate. Iodide to iodine oxidation by O_3 serves as the basis for standard analytical determinations for O_3 in air and water. The presence of trace reducing agents may account for the instability of aqueous O_3 solutions.

Ozone behaves as a simple oxidizing agent in many reactions. However, pH, accumulation of reaction products, solvent, etc., may alter the predicted reaction in aqueous solution. Typical groups that are readily oxidized by O_3 are -SH, =S, -NH_2, =NH, -OH (phenolic), and -CHO. Ozone may react with activated carbon to form explosive end products, and this fact should be kept in mind in tertiary treatment applications. Probably ozone's great ability to oxidize organics accounts for its widespread use in removing color, odor, and tastes. This same ability can and has been used to kill microorganisms in water.

The mechanism of disinfection with O_3 appears to be directly related to the following reaction:

$$O_3 \longrightarrow O_2 + O$$

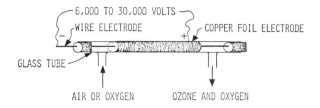

Figure 43. A Simple Laboratory Ozonator

The nascent oxygen produces a high energy oxidation via a
free radical mechanism. Bacterial cells, viewed after
ozonation by means of electronmicroscopy, appeared to have
exploded. No rigid structural features were able to with-
stand the oxidation process. Cells took on the appearance
of a splash of gelatinous material.

GENERATION OF OZONE

O$_3$ must be generated at the site where it is to be used.
It is produced usually by one of three techniques:
(1) electrical discharge, (2) electrolysis of perchloric
acid, or (3) ultraviolet lamps. Relatively high concentrations of O$_3$ are produced by the first two methods, but
the third produces only approximately 0.003 grams per hour
for each 1/100 of a watt. The only practical method of
large-scale O$_3$ production has been with electrical discharge. Figure 43 depicts a simple apparatus for the
production of O$_3$ by silent electrical discharge. Ozone
is produced according to the following reaction:

$$3O_2 \xrightarrow{\text{silent electrical discharge}} 2O_3$$

Oxygen   Ozone

All commercial ozonators operate on this principle, but
there are many different electrode configurations and
cooling arrangements. Since 85 to 95% of the electrical
energy applied is dissipated as heat, a cooling arrangement
must be provided—usually a water jacket cooled by passing
water at low temperature through it.
Ozonized air contains some nitrogen pentoxide (N$_2$O$_5$)
and also nitrous oxide (N$_2$O), which is inert. The more

moisture in the air the higher the concentration of N_2O_5.
This substance combines with moisture to form nitric acid
(HNO_3), which is highly corrosive to metal surfaces. For
this reason, as well as for increased O_3 production efficiency, air or oxygen is usually cleaned, cooled, and
dried before passage through the ozonator. This means
that drying is a part of the ozonation process. Figure
1 (p. 19) is a schematic of the laboratory apparatus for
the production of O_3 using a source of ultraviolet light.
Note that the air is passed through a drying column before
entering the 20-liter flask. Figure 2 (p. 19) is a
schematic of the equipment used to provide O_3 for treatment of secondary effluent and depicts the 1-inch diameter
column used in our sewage studies. (This column was 18
ft high, and sewage was maintained to the 14-ft level
during ozonation; note the air dryer is again a part of
the equipment package. Any moisture in the air reduces
the quantity of O_3 produced as well as contributing to
the formation of HNO_3.) A 10-minute contact time for
ozonation was used in these studies. More research is
needed to find better ways to get the O_3 into the water
or wastewater that is to be treated.

After the O_3 is produced (2% by weight concentrations
in air are claimed by commercial ozonator manufacturers[6]),
it must be distributed throughout the water to achieve
disinfection. Usually a deep contactor is employed to
permit solution of the ozone. Whiting (Indiana) used a
16-ft-deep ozonizer with the O_3 added at the bottom
through porous diffuser plates. The water passes downward as the O_3 bubbles travel upward (countercurrent
flow). A contact time of 11 minutes at design flow is
provided for O_3 mixing and diffusion to take place. A
rapid mixer is sometimes used to assist in getting the
O_3 into the water. Figure 44 depicts the Whiting Indiana
Water Treatment Plant. Note the air cleaner, drier,
compressors, and ozonators which supply the O_3 to the
ozonizing tanks. Chlorine and ammonia are fed after the
O_3 to disinfect the water.

TOXICITY OF O_3

The effects on human health of breathing O_3 can vary
from coughing and sneezing to nausea and even death, depending upon the concentration of O_3 and the duration of
exposure. The Public Health Service has set a TLV of 0.1
ppm by volume as maximum safe concentration for working

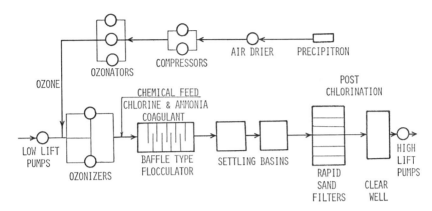

Figure 44. Flow Diagram of Whiting (Indiana) Water
           Treatment Plant

conditions.[7] Since people can begin to detect ozone's
odor at about 0.01 ppm by volume, plant operators can
easily leave the area or take corrective action when they
encounter this odor. In practice, men can do very little
useful work of the O_3 level reaches 1.0 ppm. Considerable
disagreement concerning the exact health effects attributable to specific concentrations of O_3 has resulted because
of the difficulties that have been encountered in the
accurate measurement of concentration levels.

DETERMINATION OF O_3 IN WATER

One of the problems in the use of O_3 for water disinfection has been the difficulty of accurate determination
of the concentration of O_3 obtained in the water. The
present standard quantitative method[8] for O_3 in water
involves the oxidation of KI to I_2 in the presence of
excess iodide. This method is currently used by many
workers in the environmental area. However, new methods
for determining O_3 in air and water have appeared regu-
larly in the literature. This fact suggests some limitations are connected with all of the existing methods.
The precision given for the iodometric method is ±1.0%.
This is not correct for low levels of O_3 in water. This
starch iodide procedure begins to show significant error
when the released iodine is less than about 2.0 mg/liter.

An iodine concentration of 2.0 mg/liter in neutral solution is equivalent to an ozone concentration of about 0.4 mg/liter, so the starch iodide procedure has a large error when the O_3 concentration is less than about 1.0 mg/liter. Add to this the problem of collection of O_3, the reactivity of O_3, and the usefulness of this method for sub-mg/liter concentrations of O_3 is questionable. In neutral or alkaline solutions, O_3 reacts with iodide according to the following reaction:[1]

$$H_2O + O_3 + 2I^- \longrightarrow O_2 + I_2 + 2OH^- \qquad (1)$$

In acid solutions the ratio of iodine released increases so that in concentrated hydrogen iodide solutions the following reaction occurs:

$$6\ HI + O_3 \longrightarrow 3\ H_2O + 3I_2 \qquad (2)$$

Special precautions must be taken to use the starch iodide method when working with sub-mg/liter concentrations of O_3.

Table 12 contains a summary of the methods that have been suggested for low concentrations of O_3, with the advantages and disadvantages of each indicated. We have devised a colorimetric method using the leuco base of crystal violet. This method has proved successful thus far with our research, and it is still under development for routine use.

DISINFECTION EFFICIENCY

Ozone is a powerful disinfecting agent. Numerous workers[2,16-19] have shown that relatively low concentrations of O_3—less than 0.5 mg/liter—will destroy organisms in water. Many workers have stressed the point that O_3 sterilizes the water rather than just disinfects it. Ozone concentrations of from 0.5 to 4.0 mg/liter are usually used in water disinfection applications. In these studies, O_3 concentrations as low as 0.01 mg/liter were found to be extremely toxic to *E. coli* and *S. faecalis* at pH 7.0 and 25° C and 30° C in pure systems. Higher concentrations (50 to 60 mg/liter O_3) applied to secondary effluent caused 100% destruction of fecal coliforms in from 10 to 30 minutes contact time. Total coliforms were reduced to less than 1% of their original levels by these same concentrations.

Tables 13 and 14 indicate the effect of low levels of O_3 on *E. coli* and *S. faecalis* in very pure aqueous solution

Table 12. Analytical Methods for Ozone

Method	Utilizes	Interferences or Limitations	Reference
Iodometric method	$2KI \rightarrow I_2$; titrate with reducing agent	Any oxidant interferes: NO_2, SO_2, organic matter, ferric ion	8
Orthotolidine manganese (OTM) method	Manganese to manganic ion; acid OT produces color	Nitrite, oxidized ion, semiquantitative	8
Orthotolidine arsenite (OTA) method	Ozone + OT produces color; arsenite minimizes interference	Manganese dioxide, qualitative	8
1,2-di-(4-pyridyl) ethylene	Ozonide formation which cleanses forming pyridine-4-aldehyde	SO_2, NO_2 interfere; applicable above 16° C only; perhaps measures total oxidant	9,10
Sodium diphenylamine-sulfonate	Undefined color formed measure at 593 mμ	NO_2 interferes	9
Phenolphthalein oxidation	Oxidation of colored indicator	Any oxidant interferes	11
Long path ultraviolet; long path infrared; galvanic cell	Instrumental	Lost sensitivity and costly instrumentation	12,13,14
Kinman and Layton leuco crystal violet method	Colorless leuco base of crystal violet is oxidized by O_3 to colored crystal violet	Many ions do not interfere; method under development	15

Table 13. Destruction of *E. coli* by O$_3$ in Pure H$_2$O

Run	Temp. (°C)	pH	O$_3$ (mg/liter)	Time for 100% Kill (sec)
1	25	7.0	0.01	60
2	25	7.0	0.09	20
3	25	7.0	0.10	20
4	25	7.0	0.12	15
5	25	7.0	0.19	15
6	30	7.0	0.01	(no kill)
7	30	7.0	0.05	15
8	30	7.0	0.24	15
9	30	7.0	0.27	15
10	30	7.0	0.30	15
11	30	7.0	0.31	15

Table 14. Destruction of *S. faecalis* by O$_3$ in Pure H$_2$O

Run	Temp. (°C)	pH	O$_3$ (mg/liter)	Time for 100% Kill (sec)
1	25	7.0	0.01	15
2	25	7.0	0.01	20
3	25	7.0	0.09	15
4	25	7.0	0.20	40
5	30	7.0	0.02	60
6	30	7.0	0.03	20
7*	30	7.0	0.07	20
8	30	7.0	0.08	20

*Contained 1.0 mg/liter NH$_3$ also.

at pH 7.0 and 25° C and 30° C. All water was rendered
halogen-demand free and the bacterial concentrations were
near 1X10⁶ organisms per milliliter at the beginning of
each run. Survivors were counted by dilution to BBL
Tergitol-7 agar for the *E. coli* and BBL KF agar for the
S. faecalis. A contact time of only 15 seconds was adequate to provide 100% destruction of the organisms in most
cases. Run number 6 in Table 13 was extremely interesting
and points out what happens if there is something other
than the organisms in the water to react with the O₃.
Evidently the glassware was not clean in this run. Some
organic or trace material caused the loss of the O₃ before
it could react with the organisms. In routine disinfection
applications where the water is not free of organics and
numerous trace materials, some of the applied O₃ dose will
be consumed without any destruction of organisms. Some of
these materials catalyze the rapid decomposition of the
O₃ with the result that relatively little disinfection
occurs until a so-called critical dose is attained. Some
workers call this the "all-or-none effect." This is essen-
tially an ozone-demand effect, and each water will have
its own O₃ demand just as it has a chlorine demand or
iodine demand or bromine demand. For adequate disinfection
to occur, the demand for O₃ by substances other than microorganisms must be met with a sufficient dose of O₃.

Figure 45 depicts the relative rates at which a large

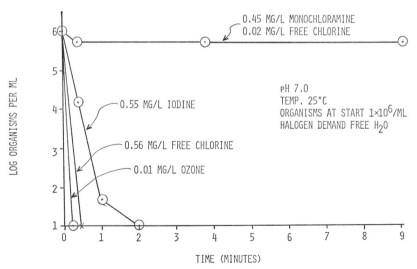

Figure 45. Destruction of *E. Coli* by Disinfectants in Water

number of *E. coli*, 1×10^6 organisms per milliliter, are destroyed by various disinfectants. Ozone appears to have the fastest rate of kill. HOCl is slower than O_3, and I_2 is somewhat slower than the HOCl but is very much faster than the monochloramine. Ozone has been shown to be more efficient against certain organisms than is chlorine. In this work, very low levels of O_3 (0.01 mg/liter) were sufficient to kill *E. coli* and *S. faecalis* in a short time. By the time a sample could be taken, 15 seconds, the organisms were destroyed. This work will be continued to determine what organisms may be resistant to O_3 and what substances interfere with or inhibit the disinfection ability of O_3.

Tables 15 and 16 contain the disinfection results[20] from 10-min ozonation of secondary effluent from the

Table 15. Destruction of Total Coliforms in Secondary Effluent

Run	COD Reduction (%)	Absorbed O_3 (mg/liter)	Time (min)	Organisms/ 100 ml	% Kill
1	32.8	61.0	0	3,200,000	0
			10	1,600	99.95
2	25.3	56.6	0	1,400,000	0
			10	730	99.95
			30	0	
3	57.2	53.1	0	2,700,000	0
			10	160	99.995
			30	0	100
4	67.5	58.6	0	567,000	0
			10	733	99.87
			20	366	99.94
5	74.8	58.0	0	467,000	0
			10	200	99.96
			40	33	99.99
6	60.8	60.5	0	400,000	0
			10	33	99.992
			40	33	99.992
7	52.2	50.5	0	1,067,000	0
			10	133	99.99
			20	66	99.994

Table 16. Destruction of Fecal Coliforms in Secondary Effluent

Run	COD Reduction (%)	Absorbed O₃ (mg/liter)	Time (min)	Organisms/ 100 ml	% Kill
1	32.8	61.0	0	300,000	0
			10	0	100
2	25.3	56.6	0	66,670	0
			10	300	99.55
			30	0	100
3	57.2	53.1	0	300,000	0
			10	66	99.978
			30	0	100
4	67.5	58.6	0	133,300	0
			10	33	99.975
			20	33	99.975
5	74.8	58.0	0	33,300	0
			10	0	100
			40	0	100
6	60.8	60.5	0	100,000	0
			10	0	100
			30	0	100
7	52.2	50.5	0	100,000	0
			10	33	99.967
			20	0	100

Cincinnati (Ohio) Mill Creek Pilot Plant (Figure 46). In each run, O_3 was applied for a 10-min period to about 1.6 liters of secondary effluent from the pilot plant. This particular sewage has a high concentration of industrial waste along with the domestic sewage. Ozone was added at the bottom of the 18-ft column through a porous diffuser. After the O_3 was added, replicate samples were taken to count the numbers of coliforms surviving. Total coliforms and fecal coliforms were determined by membrane filter techniques as described in *Standard Methods*.[8] After the 10-min ozonation period, there were coliform organisms surviving. Sometimes fecal coliforms survived and sometimes they did not. In all cases, the kill was greater

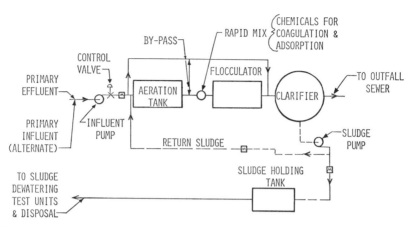

Figure 46. Mill Creek (Ohio) Pilot Plant

than 99% after the 10-min period. COD values were reduced
by from 25 to 75% of their original values. Concentrations
of O_3 greater than 50 mg O_3 per liter of sewage did not
destroy all of the coliforms in every run. This was due
to competition for the O_3 by the organics present as well
as by decomposition of the O catalyzed by the many trace
substances in the wastewater. In general, longer contact
times between the O_3 and the waste resulted in greater
bacterial kills.

Tables 17, 18, and 19 contain the results from disinfecting a water with O_3 which contains 5% raw domestic
sewage. This sewage is the same sewage used at the R.A.
Taft Center in Cincinnati for much of the research going
on there. The procedure used to disinfect this water was
as follows. Approximately two liters of demand-free
autoclaved water was buffered to pH 7.0 and brought to
contact temperature in a water bath in a brown glass bottle.
The sample of water was then ozonated to the desired level
of O_3, and a determination of initial O_3 concentration was
made. A volume of raw sewage was then added to obtain the
desired concentration of sewage by volume. The data in
Tables 17, 18, and 19 present the results of disinfection
of 5% sewage by volume, samples at pH 7.0 and 25° C, with
initial concentration of O_3 at 0.99 mg/liter, 1.50 mg/liter,
and 1.90 mg/liter.

Table 17 reveals only a 98.7% reduction in total bacteria in 15 minutes with an initial O_3 concentration of
0.99 mg/liter. When the O_3 concentration was increased

Table 17. Destruction of Total Bacteria in 5% Raw Sewage
at pH 7.0 and 25° C by O_3

Total Solids (mg/liter)	Volatile Solids (mg/liter)	O_3 (mg/liter)	Time	Organisms/ 100 ml	% Kill
555	91	0.99	0	2,500,000	0
			30 sec	133,000	94.6
			1 min	400,000	84.0
			5 min	430,000	83.8
			10 min	100,000	95.7
		0.04	15 min	33,000	98.7
545	105	1.50	0	2,500,000	0
			30 sec	33,000	98.7
			1 min	267,000	89.3
			5 min	100,000	95.7
			10 min	0	100
		0.97	15 min	0	100
463	41	1.90	0	2,500,000	0
			30 sec	333,000	86.7
			1 min	0	100
			5 min	0	100
			10 min	0	100
		0.75	15 min	0	100
		0.30	24 hrs	3,000,000	-

to 1.50 mg/liter, 100% reduction or kill occurred in 10 minutes. When the O_3 concentration was increased to 1.90 mg/liter, 100% kill was obtained in 1 minute. This is a rapid rate of kill and points out the toxicity of O_3 to a wide range of bacterial organisms. Note though that a sample taken at 24 hours provided a count of 3×10^6 organisms per 100 ml. This apparent contradiction illustrates what may happen if there are solids in the water to be disinfected with O_3, especially if these solids are of fecal origin (as was the case here) where organisms are encased in organic matter through which the O_3 could not penetrate in the time with the dispersion and concentration conditions involved. A count of 3×10^6 organisms per 100 ml resulted in the presence of an O_3 residual of 0.30 mg/liter.

Table 18. Destruction of Total Coliforms in 5% Raw Sewage
at pH 7.0 and 25° C by O_3

Total Solids (mg/liter)	Volatile Solids (mg/liter)	O_3 (mg/liter)	Time	Organisms/ 100 ml	% Kill
555	91	0.99	0	4,500,000	0
			30 sec	470,000	89.6
			1 min	230,000	95.0
			5 min	710,000	84.3
			10 min	100,000	97.8
		0.04	15 min	100,000	97.8
545	105	1.50	0	4,500,000	0
			30 sec	167,000	96.3
			1 min	6,300,000	—
			5 min	0	100
			10 min	0	100
		0.97	15 min	0	100
463	41	1.90	0	4,500,000	0
			30 sec	370,000	91.8
			1 min	167,000	96.3
			5 min	33,000	99.3
			10 min	167,000	96.3
		0.75	15 min	33,000	99.3
		0.30	24 hrs	33,000	99.3

In this instance, the mere presence of an O_3 residual did
not assure a water of sanitary quality. If there is no
turbidity present and the total solids are low, then perhaps a significant O_3 residual would indicate a water safe
to drink. However, bacteriological and virological
analyses should be run in conjunction with the analyses
for O_3 before a decision is made regarding the quality of
the water.

Table 18 contains the data from disinfecting the sewage-
laden water for members of the coliform group as determined
by plate counts on Tergitol-7 agar. Note that coliforms
remained after 15 minutes with an initial O_3 concentration
of 0.99 mg/liter. Increasing the O_3 concentration to 1.50
mg/liter in another run resulted in 100% kill in 5 minutes.

Table 19. Destruction of Fecal Streptococci Organisms in
5% Raw Sewage at pH 7.0 and 25° C by O_3

Total Solids (mg/liter)	Volatile Solids (mg/liter)	O_3 (mg/liter)	Time	Organisms/ 100 ml	% Lill
555	91	0.99	0	267,000	0
			30 sec	0	100
			1 min	0	100
			5 min	0	100
			10 min	0	100
		0.04	15 min	0	100
545	105	1.50	0	267,000	0
			30 sec	0	100
			1 min	0	100
			5 min	0	100
			10 min	0	100
		0.97	15 min	0	100
463	41	1.90	0	267,000	0
			30 sec	0	100
			1 min	0	100
			5 min	0	100
			10 min	0	100
		0.75	15 min	0	100
		0.30	24 hrs	0	100

In a third run, an initial O_3 concentration of 1.90 mg/ liter did not achieve complete disinfection. Counts of 3.3×10^4 coliform per 100 ml remained after 15 minutes and 24 hours respectively. As before, this apparent inefficiency of the O_3 results from the presence of protected organisms encased in particulate matter in the water. A significant O_3 residual would not be an assurance that this water is safe to drink. The solids would have to be removed first.

Table 19 contains the counts for organisms of the fecal streptococci group. These organisms were killed quickly and easily at each of the O_3 concentrations studied. Usually, they were all killed within 30 seconds. They did not appear to have the protection that the coliform group and the total bacteria had.

ADVANTAGES AND DISADVANTAGES OF OZONE

Table 20 lists the advantages and disadvantages of O_3 for disinfection applications. Certain factors need to be stressed. Ozone is a powerful disinfecting agent and deserves a place in our arsenal of disinfectants. In tertiary treatment applications for direct reuse of the water, where the water is known to contain viruses, O_3 may be the disinfectant of choice.

Table 20. Advantages and Disadvantages of Ozone for Water and Wastewater Disinfection

Advantages	Disadvantages
Wide spectrum disinfectant	High capital cost for equipment
Avoids taste and odor problems	Must be generated at site
Removes color	High reactivity results in low selectivity
Adds oxygen to water	Low solubility under normal conditions
Disinfection is rapid	Operation and maintenance may be a headache
Has high oxidation potential	Residual O_3 cannot be maintained in H_2O for long time periods
Lowers BOD and COD values	More expensive than chlorine at present
Low concentrations are adequate	
Does not form noxious compounds in treated water	
Avoids problems associated with transportation of potentially harmful chemicals	

Manufacturers of O_3 generating equipment predict a decrease in the cost of generation O_3 because of advances in the design of their equipment. Anyone considering O_3 should run a cost analysis of the cost of O_3 disinfection versus the cost of chlorine or iodine disinfection before they select the method of choice. Many different costs are given in the literature for O_3 disinfection. Some of the costs may be misleading unless a careful analysis is made of the basis for the costs.

Research is needed to find better ways of getting the O_3 into the water or wastewater in question. Considerable research is under way with O_3, and new dosing schemes have been proposed. These involve step additions of ozone, high shear contacting, and recycling of the ozonated air stream. Little or no full-scale cost data are available for these devices. Laboratory and pilot plant cost data may be very misleading.

Research is also needed to better define the conditions for maximum destruction of organisms in water or wastewater by O_3. If dose levels are reduced overall and O_3 production costs go down, O_3 could become more competitive in this country.

SUMMARY

Ozone has been shown to be a powerful disinfecting agent for either water or wastewater applications. A significant residual ozone concentration does not guarantee that a water is safe to drink. Organic solids may protect organisms from the disinfecting action of ozone and increase the demand for ozone. Ozone residuals cannot be maintained in metallic distribution systems for very long. The inability to maintain an ozone residual in water distribution systems is the principal reason ozone is not more widely used for water disinfection in this country and the principal reason for its use as a pretreatment in those water treatment plants where it is in use.

ACKNOWLEDGMENTS

This work was partially supported by U.S.P.H.S. grant No. 5-P10-ES-00159-03 to the Institute of Environmental Health and F.W.Q.A. Training Grant No. 5T1-WP-209-02 to Dr. J. D. Eye. This support was greatly appreciated.

The author wishes to acknowledge the assistance of Arthur D. Caster and members of his staff at the Mill Creek Sewage Treatment Plant of the M.S.D. of Greater Cincinnati. Their help and use of their facilities were greatly appreciated.

Special thanks are in order to Janet Rickabaugh and Theresia Mesgetz for help in the laboratory and to June Schuck for typing this manuscript.

REFERENCES

1. Hann, V. A., and Manley, T. C., "Ozone." Encyclopedia of Chemical Technology, *9*, 735 (1952).
2. Bartuska, J. F., "Ozonation at Whiting, Indiana." J.A.W.W.A., *33*, 2035 (1941).
3. Bean, E. L., "Ozone Production and Cost." Advan. Chem. Ser., *21*, 430 (1956).
4. Whitson, M. T., "The Treatment of Water with Ozone." Manchester and District Assn. of Inst. C. E., presented March 20, 1940.
5. Hann, V., "Disinfection of Drinking Water with Ozone." J.A.W.W.A., *48*, 1316 (1956).
6. ———, "Ozone Bids for Tertiary Treatment." Env. Sci. and Tech., *4*, 893 (1970).
7. M.S.A. Technical Information, Section 10, Mine Safety Appliances Company, Pittsburgh, Penn.
8. American Public Health Assn., Standard Methods for the Examination of Water and Wastewater, 1963-1966. A.P.H.A., New York (1966), p. 219.
9. Hauser, T. R., and Bradley, D. W., "Specific Spectrophotometric Determination of Ozone in the Atmosphere using 1,2,-Di-(4-Pyridyl) Etheylene." Analytical Chemistry, *38*, 11 (1966).
10. Hauser, T. R., and Bradley, D. W., "Effect of Interfering Substances and Prolonged Sampling on the 1,2-Di-(4-Pyridyl) Etheylene Method for Determination of Ozone in Air." Analytical Chemistry, *39*, 10 (1967).
11. Deutsch, S., A.P.C.A., *18*, 78 (1967).
12. Haagen-Smit, A. J. and Brunelle, M. F., Air Pollution, *1*, 51 (1958).
13. Hanst, P. L., et al., "Absorptivities for the Infrared Determination of Trace Amounts of Ozone." Analytical Chemistry, *33*, 8 (1961).
14. Hersch, P., and Derringer, R., "Galvanic Monitoring of Ozone in Air." Analytical Chemistry, *34*, 7 (1963).

15. Kinman, D. N., and Layton, R. F., "A New Analytical Method for Ozone." Proceedings, ASCE San. Div. National Conference on Disinfection, Amherst, Massachusetts, July 1970.
16. Lebout, H., "Fifty Years of Ozonation at Nice." <u>Advan. Chem. Ser.</u> *21*, 450 (1959).
17. Torricelli, A., "Drinking Water Purification." <u>Advan. Chem. Ser.</u>, *21*, 453 (1959).
18. Whitson, M. T., "Symposium on the Sterilization of Water, (D) Other Processes with Special Reference to Ozone." <u>J. Inst. Wtr. Engrs.</u> (Br.), *4*, 600 (1950).
19. O'Donovan, D. C., "Treatment with Ozone." <u>J.A.W.W.A.</u>, *57*, 1167 (1965).
20. Roush, P. H., "Ozone Disinfection of a Combined Industrial-Domestic Wastewater." Unpublished thesis, Univ. of Cincinnati, Cincinnati, Ohio (1970).
21. The Welsbach Corporation, "Welsbach Ozone." Ozone Processes Div., Philadelphia, Pennsylvania.
22. Bartuska, J. F., "Ozonation at Whiting: 26 Years Later." <u>Public Works</u>, August 1967.
23. Evans III, F. L., and Ryckman, D. W., "Ozonated Treatment of Wastes Containing ABS." Presented at Purdue Industrial Waste Conference, Purdue Univ., Lafayette, Ind., May 1963.
24. Frison, P., "Development of European Ozonation Techniques." <u>Advan. Chem. Ser.</u>, *21*, 443 (1959).
25. Huibers, D. Th. A., et al., "Ozone Treatment of Secondary Effluent from Waste Water Treatment Plants." Robert A. Taft Research Center, Report No. TWRC-4, FWPCA, April 1969.
26. Diaper, E. W. J., "A New Method of Treatment for Surface Water Supplies." <u>Water and Sewage Works</u>, *117*, 373 (1970).
27. Gomella, C., "Ozone Practices in France." Presented at AWWA Annual Meeting, Denver, Colorado, June 13-18, 1971.
28. Sommerville, R. C., and Rempel, G., "Ozone for Supplementary Water Treatment." Presented at the AWWA Annual Meeting, Denver, Colorado, June 13-18, 1971.

CHAPTER VIII

PRACTICAL ASPECTS OF WATER AND

WASTE WATER TREATMENT BY OZONE

E. W. J. Diaper

INTRODUCTION

Ozone has been used for the disinfection of water
supplies since the beginning of the century when it was
first applied to the treatment of water for the City of
Paris, France. There are now nearly 1000 installations
in operation, mainly in Europe but also including 20 in
Canada, where the largest is operating on drinking water
supplied to the City of Quebec, treating flow rates up to
60 mgd.

In this country, applications of ozone for water treatment are not common because of the widespread use of
chlorine, which is readily available and can be easily
and economically used for disinfection. As it is not possible to bottle ozone, installation costs are normally
higher than with chlorine, and, until recently, running
costs also were generally higher.

With the advent of modern ozone generators, resulting
in reduced installation and operating costs, and with the
increased emphasis on pollution control, ozone is receiving
renewed attention. New techniques using ozone have been
developed for the treatment of water supplies and effluents.
Ozone has many desirable features for water and waste treatment, among these being its powerful oxidizing properties,
almost instantaneous action, and absence of a permanent
residual.

This chapter provides an opportunity for reviewing recent progress in the design of ozone generators and applications of ozone for water and wastewater treatment.

CHEMICAL NATURE OF OZONE

If oxygen or air is passed through an electric discharge or exposed to certain wave lengths in the ultraviolet range, some of the oxygen is polymerized and ozone is produced following the reaction:

$$3 \ O_2 = 2 \ O_3 - 69 \text{ kc (or } 34.5 \text{ kc per mole)}$$

It can be imagined that oxygen molecules represented by O_2 are dissociated into atomic oxygen represented by $O + O$. Double and triple collisions of the O atoms can form new molecules, some O_2 and others O_3—ozone. Additionally, collisions between atoms and molecules of oxygen may also produce O_3. Some ozone molecules so formed can also be broken up. By the laws of probability, the mechanism of production allows only a portion of the oxygen exposed in the gas flow to be converted to ozone.

Oxygen gas is colorless, odorless, tasteless, and non-toxic, with a density of 1.42904 gm/liter at normal temperature and pressure, or 1.105 when air equals 1. It is soluble in distilled water up to 8.92 mg/liter at 20° C and 12.3 mg/liter at 5° C.

Ozone has considerably different properties from oxygen. It has a characteristic odor, thought by some to be re-freshing in low concentrations. In high concentrations the gas is toxic, although by nature of its production it is less hazardous for practical use than many bottled gases.

Density of ozone is 1.5 times that of oxygen, and it is some 10 times more soluble in water. The attachment of the oxygen atom to the oxygen molecule is unstable, and ozone will eventually revert spontaneously to oxygen. Ozone is more rapidly decomposed by heat than is oxygen; at a temperature of around 270° C (518° F), it is immedi-ately converted back into oxygen.

Because of the inherent instability of its molecular structure even in low concentrations, ozone is capable of producing a series of virtually instantaneous reactions in liquid or gaseous phase contact with oxidizable sub-stances. The nature of these may be illustrated by the liberation of iodine from a solution of potassium iodide, which is a reaction employed for quantitative estimation of ozone.

$$O_3 + 2\ KI + H_2O = I_2 + 2\ KOH + O_2$$

The active element is atomic oxygen released from the ozone molecule which reverts to molecular oxygen.

PRODUCTION OF OZONE

Of the various means by which ozone can be generated, electrical production is the only practical and economical method for large-scale use. The process involves the passage of air (or oxygen) between electrodes across which an alternating high-voltage potential is maintained. To ensure the conversion of an optimum part of the oxygen into ozone, a uniform blue-violet glow discharge is maintained throughout the gas. The glow discharge is created by inserting a dielectric material between the electrodes, which causes the glow to spread uniformly and prevents breakdown into brush and arc discharge.

Air passing through the electrodes must be effectively filtered to ensure the absence of dust. In principle, cool air is desirable, but temperature has very little effect because the air, with its low specific heat, will quickly acquire the temperature of the electrode area.

The air must be dry. Drawn from the atmosphere, air contains a highly variable proportion of water vapor. In extremely hot saturated ambient conditions, as much as 5% by weight can be held as vapor (at 40° C). For efficient production of ozone, the water in the air must first be removed so that the dew point is not higher than -50° C (-60° F). With water vapor concentrations higher than this, not only is ozone production affected, but also oxides of nitrogen are produced which accelerates the decomposition of ozone as well as causing corrosion of metals. Risk of breakage of dielectrics is also increased. The effect of dew point on ozone production is shown in Figure 47.

There is no economic advantage in reducing the dew point below -50° C; at lower values, the additional production of ozone does not compensate for the extra cost of air conditioning. A dew point of -50 ° C can be reached by silica gel or activated alumina desiccation alone, which is often used in small installations. In larger plants, desiccation preceded by refrigeration is desirable. A temperature near 0° C at inlet provides the most favorable condition for the efficient operation of the desiccator (maximum relative humidity, low temperature).

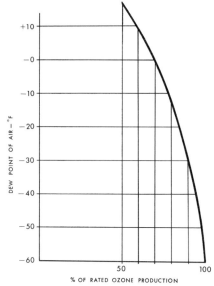

Figure 47. Influence of Dew Point on Production of Ozone

There is also an economic advantage in refrigeration since each gram of water removed by cooling requires a power consumption between .3 and .35 watts per hour, while the removal of a similar amount of moisture by desiccation needs approximately 7 to 10 times this energy. In summer conditions with air carrying between 15 and 20 grams of water per cubic meter, more than half can be removed economically by refrigeration. It does not pay to go much further. To reach the low dew points required for ozone production by refrigeration alone is not economical.

Desiccation between refrigeration and ozone production gives security in the event of refrigeration breakdown, ensuring the supply of dry air at all times even though the desiccator regeneration cycle may be affected. An additional advantage of operating the refrigeration stage to give only a moderately low dew point is that the working temperature is maintained between 0 and 5° C. Frosting does not occur and duplex coils are unnecessary.

When using oxygen for the production of ozone, refrigeration and desiccation for drying purposes are normally not required. This is because oxygen is produced at low temperatures and is initially dry. However, if recycling is used to economize on the use of oxygen, then drying equipment similar to that used for air preparation would be required.

Electricity for ozone generation can be taken from the normal main supply. Because of the electrical capacitance of the Ozonator, the frequency that can be employed has certain limits. Within these limits, the power absorbed increases with frequency, and ozone production is nearly in proportion to the power, efficiency remaining about constant, provided that optimum adjustment of the discharge gap is obtained.

Standard AC frequencies of 50 or 60 cycles are suitable. Five hundred cycles is also employed, and, in some cases, 1000 cycles. Transformers are used to step up the line voltage to operating values.

There are advantages in keeping the high-tension voltages moderate, consistent with efficient ozone production. The usual range of voltage is between 5 and 20 kilovolts. A maximum of 20 kilovolts is usual. Above this, damage to glass dielectrics becomes an increasing risk, requiring the use of more expensive materials. Problems of transformer design and of cable and terminal insulation are also minimized by restricting voltage.

To produce ozone, the dry air is passed between pairs of high-tension and low-tension (earth) electrodes, the latter shielded by dielectric material with a critical gap between the dielectric and the high-tension electrode. This gap usually varies between 1 and 3 millimeters, depending upon power and air flow and thus upon O_3 concentration. The glow discharge is maintained in this gap. The threshold voltage for glow discharge, a function of the nature of the dielectric, increases with dielectric thickness, which also affects the voltage necessary to obtain a given power.

Heat as well as ozone is generated, and the system must be cooled to safeguard the materials of construction—especially the dielectric—and to prevent destruction of the ozone as it is formed. Thin dielectrics are an advantage in cooling, especially as they are invariably poor conductors of heat. Glass and Pyrex between 1 and 3 millimeters thick is often used as the dielectric material.

Directly controllable variables in ozone production are the rate of air flow and voltage across the electrodes. Other variables, indirectly controllable, are air, humidity, and temperature. Performance curves for a typical ozone production unit are shown in Figure 48. Illustrated are the relationships existing among five factors: air flow, total power input, power per gram of ozone produced, ozone concentration in air, and total ozone production.

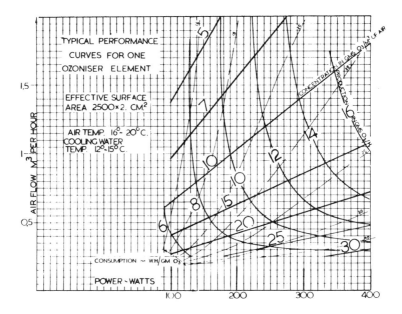

Figure 48. Typical Performance Curves for One Ozonizer
Element

In water and waste treatment, total ozone production
is not the only criterion. Ozone concentration in air is
also important. Although there is an economic interest in
employing low concentrations, it is the high ozone concentrations that usually give the best results.

Using air, it is economically possible to obtain concentrations of ozone up to 25 grams per cubic meter
(approximately 2% by weight). Employing oxygen and depending upon operating conditions, the quantity of ozone
and concentration may be increased 2 or 3 times. Power
consumption for a given ozone concentration, using oxygen,
is about half that required using air.

OZONE GENERATORS

Ozone generators can be divided into two main classes:
(1) Ozonators with tubular electrodes, and (2) Ozonators
with flat electrodes.

Horizontal Tubular Ozonator

The horizontal tubular Ozonator is a cylindrical, water-tight vessel, usually made of stainless steel, containing horizontal metal tubes fixed at their ends into two tube sheets which divide the vessel into three compartments. The middle compartment allows a circulation of water for cooling the tube assemblies. One of the end compartments receives the dry gas, air, or oxygen, and the other collects the ozonized gas after its passage through the tubes.

Each metallic tube constitutes a ground electrode, inside of which is arranged a cylindrical coaxial tube of glass acting as a dielectric whose internal metallized surface is connected to the high voltage. The dielectric tube is sealed at one end and centered by means of stainless steel rings. Its exterior diameter, slightly less. than the interior diameter of the ground tube, gives an annular space in which the glow discharge is established and through which the gas to be ozonized passes. This Ozonator is capable of supporting a small pressure, 15 psi maximum, which may be used to introduce the ozone into its point of use. The latest tubular element apparatus can produce more than 15 grams per hour per tube at a concentration of 20 grams of ozone per cubic meter of air with specific consumption of energy of 17 watts hours per gram. These results relate to the supply of electrical energy at a frequency of 50 cycles. They can be improved by using a supply of higher frequency. The largest generators of this type include more than 500 tubes and produce 7500 grams per hour (390 lbs/day). They are used in the large waterworks installations in Paris, France (Figure 49).

Vertical Tubular Ozonator

The discharge elements in this tubular Ozonator are arranged vertically in a cylindrical vessel having three chambers or compartments as in the first type described. The upper chamber receives and distributes the dry gas through a group of metallic tubes which constitutes the high-voltage electrodes. Each metallic tube is arranged coaxially in an insulating tube of special glass, closed at the lower end, and with an interior diameter slightly larger than the exterior diameter of the metallic tube, providing an annular space for ascent of the gas.

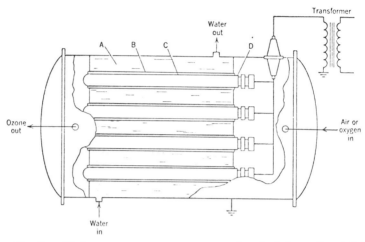

A - Cooling Water;   B - Stainless Steel Tube;
C - Discharge Gap;   D - Glass Tube.

Figure 49. Horizontal Tube Ozonator

    In the interior compartment of the vessel, which acts
as a ground electrode, the dielectric tubes are immersed
in cooling water over almost their entire length. The
glow discharge is established in the annular space between
the dielectric and the high-tension electrode. The ends
of the dielectric tubes open into a second chamber situated
between the upper and lower compartments which collect
the ozonized gas after its passage through the tubes.
    This type of generator will also allow operation under
pressure. Outputs at 50 cycles and at several hundred
cycles per second are given in Table 21.

Vertical Plate Ozonator

    First made by Otto in the year 1906, the vertical plate
Ozonator is now included in more than 500 installations,
the great majority presently in service. Both electrodes
and dielectrics are flat with a central orifice. The
earth electrode is a hollow aluminum alloy plate having
perfectly flat and parallel external faces through which
the cooling water circulates. The dielectric is a flat

Table 21. Production Characteristics—Vertical Tube
Ozonator

Parameter	Ozone Output per Tubular Element			
	f = 50 cps		f = several hundred cps	
Concentration, g/cu m	15	20	15	20
Concentration, % by wt.	1.25	1.67	1.25	1.67
Power, watts	320	320		
Output of ozone, gm/hr	19.9	18.0	67.1	49.6
Output of ozone, lbs/day	1.05	0.95	3.55	2.62
Specific power, Wh/gm	16.1	17.7		
Specific power, Kwh/lb	7.3	8.05		

sheet of glass set against the earth electrode. The high-tension electrode is a thin stainless steel plate. Air or oxygen flow takes place in the space between the high-tension electrode and the dielectric.

Electrodes and dielectrics are placed successively. When placed close together, they form a block with central orifices which make up a channel through which ozone flow takes place.

The gas to be ozonized is admitted around the outside of the assembly, under approximate atmospheric pressure, and flows across each plate to the central channel, where it is collected and evacuated under a small negative pressure.

The capacity of the ozone generator depends upon the number of blocks. Cooling water flows through the low tension electrodes from two manifolds, one for admission, the other for collection, with takeoff points to each plate. Units of this type incorporate 84 elements having an output at 50 cycles of 2.1 to 2.3 kilograms per hour (109-120 lbs/day) and production at a higher frequency of 5.1 to 6.3 kilograms per hour (265-328 lbs/day) with concentrations between 20 and 15 grams per cubic meter. (See Table 22.)

Table 22. Production Characteristics—Vertical Plate
Ozonizer

Parameter	Ozone Output per Vertical Plate Element			
	f = 50 cps		f = several hundred cps	
Concentration, g/cu m	15	20	15	20
Concentration, % by wt.	1.25	1.67	1.25	1.67
Power, watts	500	500	1175	980
Output of ozone, gm/hr	27.6	24.9	75.2	60.6
Output of ozone, lbs/day	1.46	1.31	3.98	3.2
Specific power, Wh/gm	18.1	20.1	15.6	16.1
Specific power, Kwh/lb	8.2	9.15	7.1	7.3

Horizontal Plate Ozonator

In this unit, the electrodes are not square, with a
central orifice and hung in a common cabinet as they are
in the vertical plate Ozonator; instead, each element is
a self-contained block with horizontal rectangular electrodes.   An Ozonator is made up of the required number
of elements, stacked one above the other.

The upper and lower sections of each block are spaced
well apart and form the low-tension electrodes.   The two
metal plates (comprising the high-tension electrodes) and
the glass dielectric plates are held tightly inside the
block, with the glass dielectric against the low-tension
electrode.   The block is sealed at each end by rectangular
covers.   The resulting element has an exceptionally long,
well-defined discharge area, with one end open to receive
dry air and the other fitted to a pipe outlet for ozonized
air discharge.   On one of the longitudinal sides of the
block are pipe fittings for inlet and outlet cooling water
connections to the low tension-electrodes and for outlet
of ozonized air.

Each block has an individually fused high-tension
electrical supply and is fitted with a porthole allowing
inspection of the electrodes, dielectrics, and discharge
area.

An assembly of a number of blocks stacked vertically comprises an Ozonator, with additional parts to hold the blocks rigidly together. This arrangement needs only a single entry of dry air for the Oaonator. The electrical, cooling water, and ozonized air connections for each block are gathered on one side of the Ozonator, forming a service area which can be adjacent to another assembly of blocks with a central service area for two stacks.

Production of ozone from a single element with 50-cycle electrical supply at atmospheric pressure, and under a small pressure (approximately 8 psi) is shown in Table 23.

Table 23. Production Characteristics—Horizontal Plate Ozonizer

Parameter	Ozone Output at 50 cps per Horizontal Plate Element			
	Atmospheric Pressure		Pressure at 8 psi	
Concentration, g/cu m	15	20	15	20
Concentration, % by wt.	1.25	1.67	1.25	1.67
Power, watts	705	705	1300	1300
Output of ozone, gm/hr	44.8	38.7	68.4	60.3
Output of ozone, lbs/day	2.36	2.05	3.62	3.18
Specific power, Wh/gm	15.7	18.2	19.0	21.5
Specific power, Kwh/lb	7.15	8.25	8.65	9.8

HIGH TENSION TRANSFORMERS

The electrical transformer for ozone production is a single-phase unit, normally suitable for operation at either 50 or 60 cycles. Since the loading is capacitative, the power factor for the combination of transformer and Ozonator is leading and has a value between 0.5 and 0.7. In a multiple installation fed by a three-phase supply, the transformers would be connected across phases so as to balance the load.

Power output is controlled by tappings on the primary windings arranged to give secondary voltage over the

required range. The secondary windings have three tappings
giving -5%, 0, and +5% of the secondary voltage to enable
optimum settings to be obtained during commissioning of
the plant. The selected tapping is connected through a
three-way remote-controlled no-volt commutator to an ade-
quetely finned porcelain insulated high tension terminal
carried on the transformer casing.

The power transformer must be specially designed, taking
account of its resistance and that of the Ozonator in order
to produce resonance while avoiding a related excess of
voltage. The transformer is supplied either at 50 or 60
cycles or at higher frequencies of 150 to 1000 cycles,
using either an alternator for frequencies not exceeding
500 or a static frequency change.

Alternators had been used for more than 30 years, but
their high cost and low output, which impairs the overall
efficiency, have caused them to be abandoned. At the same
time, the considerable advances achieved on an Ozonator
supplied at 50- or 60-cycles frequency have made their use
unnecessary. For the same reason, static frequency changes
have been only little used until now, but appear to have
an advantage at medium frequency, particularly in regard
to capital costs for large installations.

The absorbed power, on which depends the production of
ozone and which is the function of the secondary voltage
at a given frequency, is controlled by different procedures.
First, by switching of the primary stage on the transformer;
second, by using a subsidiary autotransformer on the primary
stage; third, variable self-saturation; fourth, voltage
change by inductance; and so on. Some of these procedures
allow a continuous variation of the voltage, and thus of
the power and ozone production, and lend themselves particularly well to control of ozone in relation to the water
flow or control from the residual of ozone in the water.

MATERIALS USED IN OZONATOR CONSTRUCTION

Most of the equipment in ozonation installations is of
standard manufacture employing construction materials of
good engineering practice. Special materials are required
principally for parts in contact with ozone.

Injection equipment, pipe work in contact with ozonized
air, and air-water emulsions can be of stainless steel or
PVC. Electrodes should be constructed of metals with good
heat conductivity and inert to dry ozone. Aluminum alloys
and stainless steels are suitable. In the low-tension

electrodes, attention must be paid to surfaces in contact with cooling water. Protective bitumen coatings are sometimes used.

Dielectric plates can be made from ordinary borosilicate glass. Other insulating and jointing fittings, as well as cooling water piping within the ozone cages, can be made from PVC. Electrical fittings and suspensions in contact with ozone can be chromium plated brass or bronze. For the ozonizer cubicles, mild steel protected by epoxy resin paint or stainless steel are suitable. Valves carrying ozonized air can be made of metal coated with ozone-resistant materials, or they can be manufactured entirely of PVC. No special requirements are necessary for concrete in contact with ozonized water.

PLANT EQUIPMENT AND INSTRUMENTATION FOR CONTROL

The equipment is divided into three main sections, dealing with air preparation, ozone production, and ozone injection.

In a large-scale plant, local and remote control and indication are normally required, with automation as necessary. Air preparation is invariably controlled automatically. Ozone production is related to the flow of water under treatment and to its dosage requirements, both of which may vary. Production may be controlled manually or automatically by probe and relay, or by pattern. Starting up sequences are important and require the establishment of exact procedures and timing. Once these factors have been established for a particular plant, they are strictly repeatable, which makes possible automatic control of startup by electric actuation under the control of time-delay relays. All initiating controls and relevant indications with tell-tales, interlocks, and other safeguards can be assembled on a central control desk or panel and, if necessary, repeated at local positions and information centers. In this respect, the possibilities are similar to the control arrangements of pumping and power stations.

As far as instrumentation is concerned, the following summarizes the special measuring equipment required.

Water Flow: Meters for flow into contact columns (a total flow meter is also desirable).

Air Flow: Meters for the measurement of flow of ozonized air.

Water Pressure: Indicators for the operating supply.
Air Pressure: Indicators for air preparation systems.
Air Temperature: Temperature indicators for air at
inlet and outlet of desiccators.
Humidity: Indicator with alarm system for measurement
of air leaving desiccators.
Measurement of Ozone Residual in Water: Ozone meters
for indicating residual ozone after treatment.

ELECTRICAL CONTROL EQUIPMENT AND INSTRUMENTATION

Electrical equipment is largely of standard type, but
certain specialized components are required. The assembly
is designed according to the type and degree of automation
and remote control required. The total may be summarized
as follows.

Transformers and Voltage Regulation: For ozone production control, special characteristic transformers
with primary circuit tappings to vary the high tension
applied to the electrode system in the ozone generators.
Switching: The necessary contactor and circuit breaking
equipment for all electrically operated units giving
means of control and protection actuated through relays
and pushbuttons and indicated by pilot lights. Panel
and desk mounting according to requirements.
Measuring Instruments: Volt meters, watt meters and
ammeters for the various units of standard types.
Cables: Low- and high-tension cables of standard types.

DESIGN AND CONSTRUCTION

The total plant for treating a water supply with ozone
comprises the following equipment in units or multiple
units as necessary for the total flow capacity.

Air filter
Air blower
Air refrigerator
Air desiccator
Air flow pressure and humidity meters
Thermometers
High-tension transformer
Voltage regulator (in certain cases)
Ozone generator
Ozone injection equipment

Water flow meters
Pressure gauges
Ozone meters
Electrical control panels, divided for large installa-
    tions into switch cubicles and control desks
Hydraulic works, including pipes, channels, tanks,
    sluices, and valves
Other constructional work, including housing

A typical layout for the production of ozone and its
injection into water is shown diagramatically in Figure
50.

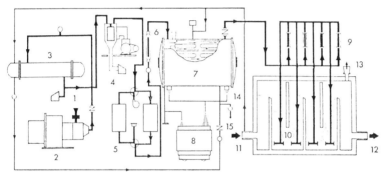

1	AIR INLET	6	AIR FLOW MEASUREMENT	11	INLET OZONISED-AIR-WATER EMULSIFICATION TANK
2	ROTARY AIR COMPRESSOR	7	OZONISER	12	OUTLET OZONISED-AIR-WATER EMULSIFICATION
3	AIR COOLER	8	H.T. TRANSFORMER	13	AIR RETURN TO ATMOSPHERE
4	REFRIGERATOR	9	OZONISED-AIR MEASUREMENT	14	COOLING WATER SUPPLY
5	AIR DRIER	10	POROUS DIFFUSERS	15	COOLING WATER DISCHARGE

Figure 50. Ozonation Plant Schematic Arrangement

METHODS OF DOSING WATER WITH OZONE

When chlorine gas is employed, its injection presents
no real technical problem. An aqueous solution is readily
made and dispersed in a relatively small flow throughout
the flow of water undergoing treatment.

Ozonation requires a somewhat more elaborate technique.
A mixture of ozone and air or oxygen cannot, without difficulty, be made into a small flow of solution for dispersal
in the main water flow. It is better to form an emulsion
of ozonized air and water, to allow a certain contact time,

and then to release to atmosphere the surplus air from
which ozone has been extracted in the water.

Some ozone is dissociated merely by coming into contact
with water and water vapor (this can be shown by ozonation
of triple distilled water); some will react with any
oxidizable substances present. This action can be virtually
instantaneous, as may be seen by the sudden reduction of
organic color at the point of ozone injection. A further
part goes into solution, and some may persist in the gas
bubbles and escape unchanged with the surplus air.

Since the object of ozonation is to secure the maximum
oxidizing effect with the minimum dose, conditions should
be created for maximum activity across the gas/water
interfaces. These conditions are secured by dividing the
ozonized air into the largest possible number of the
smallest possible bubbles as it mixes with the water,
while maintaining relative motion between the bubbles and
the water. There are various ways in which an air/water
emulsion can be produced, all requiring the expenditure
of energy. Methods include porous diffusers, jet pumps
working on the venturi principal (usually known as injectors), and the emulsion turbine.

PRINCIPAL METHODS OF INJECTION

Porous Diffusers

A feature in the use of porous diffusers is that they
involve no loss of hydraulic head in the flow of water
under treatment. The necessary energy is obtained from
compression of the ozonized air which is released through
the diffuser located at the base of the contact column.
The air bubbles created by the diffuser rise to the free
surface against the downward flow of water. A suitable
air pressure for a water depth of 15 ft, allowing for
losses in the air pipe work and in the diffuser itself,
might be 20 ft of water. Power requirements could be 28
watt hours per cubic meter of air introduced, employing
special high-efficiency compressors.

Advantages of the diffuser system, apart from the ab-
sense of hydraulic losses in the water flow, include the
ease with which heavy ozone doses can be applied without
hydraulic complications and the facility with which air
flow can be increased or decreased by simple valve control.

Ozone losses can be high unless special precuations
are taken, such as fractionizing the contract columns and
partially recycling the air leaving the columns.

Injectors

The two systems are described as partial and total
injection, terms which, although recognized, are not
properly descriptive. In both, the ozonized air forms an
emulsion with the total flow of water. The energy is
obtained from a water jet: in one case, a small proportion of the total water flow (motive flow) at high pressure
(partial injection); in the other, the total water flow
at low pressure (total injection). In the partial system,
the motive flow would normally be pumped; in the total
system, gravity head preferably would be employed. In
both cases, a free water surface in the contact tank is
necessary for release of surplus air to atmosphere. It
must be noted that, because of the possibility of surplus
air carrying traces of ozone, ventilation should be to a
safe disposal point, either by a natural draft or by forced
ventilation. If this is impossible in any particular
situation, surplus ozone can be destroyed by passing the
air through a heater and raising the temperature to about
270° C, at which point ozone is immediately dissociated.
A simple oil-fired furnace can be used for this purpose.

Emulsion Turbines

Water to be ozonated flows under gravity into the
mixing compartment, where a stirring action is developed
by a turbine attached to an electrically driven rotor.
Ozone enters below the turbine and flows outward through
small holes in the circumference of the rotor. The flow
of ozone is created by a small negative pressure induced
by the movement of the rotor, assisted, if necessary, by
a blower. As ozone passes through the holes in the rotor,
it meets the water, and the rapid shearing action at the
interface creates a fine emulsion and good mixing.

The mixture of ozone and water passes out of the mixing
compartment into the contact chamber, which is baffled to
prevent short circuiting. The contact chamber is sealed
at the top with a vent for exhaust air. The turbine rotor
is slightly tapered to assist in mixing of the ozonized
air and water.

Advantages of the system are the gravity flow conditions with virtually no head loss in the water. The dose

of ozone is independent of the water flow, and good mixing
is obtained with fine bubbles and violent shearing action.
Disadvantages include the power needed to drive the rotor
at speeds up to 3000 rpm, which can be as high as 100 watt
hr/cu meter of inspired air; the materials of construction
are expensive, since the rotor must be made of stainless
steel; and losses of applied ozone may be as high as 30%.

APPLICATIONS OF OZONE

Municipal Water Treatment—Disinfection of Water Supplies

This use of ozone has been continuous for nearly 70
years, first in France, then extending to Germany, Holland,
Switzerland, and other European countries, and most recently
to Canada.

Many investigations have been made on the relative power
of ozone and chlorine in the destruction of bacteria and
viruses. According to Bringman,[1] destruction of bacteria
in test conditions was between 600 and 3,000 times more
rapid by ozone than by chlorine. Fetner and Ingols[2] reported
that the bactericidal action of chlorine is progressive,
while that of ozone is sudden and total after a "threshold"
is the rate of dosage.

A report by Kessel et al.[3] showed that similar dilutions
of polio virus treated with chlorine to a residual of 0.5
to 1 mg/liter and with ozone to a residual of 0.045 to
0.45 mg/liter were rendered inert by the chlorine in 1-1/2
to 3 hours and by the ozone in only 2 minutes.

At the 45th General Meeting of the Society of American
Bacteriologists, a report[4] was presented indicating that
the bactericidal action of ozone was relatively unaffected
by pH values, whereas that of chlorine was considerably
affected by changes in pH value. Comparatively low concentrations of ozone were shown to bring about rapid destruction of polio virus and cysts of Entamoeba, whereas chlorine
at higher concentrations was ineffective.

Ridenour and Ingols[5] showed that ozone rendered a
dilution of 1 to 500 virus inactive in 4 minutes.

In comparing ozone with chlorine, the objection is
sometimes raised that it is not possible to maintain an
ozone residual in the water for more than a very short
period. Chlorine, on the other hand, can give persistent
residuals. Commenting on this, Whitson[6] observes of the
installation at Ashton, England, that no deterioration in
the quality of the water has been noted in the sections

supplied with ozonized water over a period of 30 years and that, in any case, the small residual amount of chlorine would have no real effect in the event of a major accidental pollution within the distribution system.

In practice, an ozone residual is established, measured 5 minutes after injection at outlet from the contact column to ensure complete sterilization. This residual is maintained at 0.1 to 0.2 mg/liter for the destruction of bacteria and 0.4 mg/liter for the destruction of viruses. The dosage of ozone applied to obtain these residuals would normally be between 0.5 and 1.5 mg/liter.

Taste and Odor Control

This provides a specific application and often justifies the use of ozone for sterilization also. Ozone will deal effectively with tastes and odors, which are accentuated when chlorine is used. The effect of ozone on chlorophenol tastes appears to be total with modest doses of between 1 and 2 mg/liter, which, in any case, would be used to effect sterilization. A great advantage is seen here by comparison with chlorine which, in any reasonable dosage, is likely to accent chlorophenol troubles. It has been reported[7] that total destruction of phenol by ozone requires a dose of between 1.19 and 1.32 mg/liter to deal with 0.2 mg/liter of phenol.

Another benefit of ozonation is the aeration effect, since the quantity of air entering the water during ozonation is appreciable and the process therefore introduces appreciable aeration as well as providing a strong oxidant. Aeration was one of the first means used in attempting to reduce taste and odor. It is said to remove hydrogen sulfide and to reduce odors due to decomposing organic matter. It appreciably reduces the carbon dioxide content with the side effect of reducing corrosiveness.

Elimination of Iron and Manganese

Ozone is employed for this purpose in water treatment, the oxidation of both iron and manganese being carried out before coagulation. With underground waters not otherwise requiring filtration, ozone followed by catalyst filtration can be employed for the elimination of these metals. An outstanding example using activated carbon filters as catalyst is at the Dusseldorf Water Works in Western Germany for a flow capacity of 38.5 mgd.

Iron and manganese are also eliminated in the MD process described below. The dose of ozone required for their elimination depends on a number of factors. In the case of Dusseldorf, about 1.5 mg/liter is required.

Removal of Color

Ozone is most effective in removing organic color. In water treatment, the yellow-brown color due to humic substances in dissolved or colloidal form is oxidized to leave the water sparkling and blue in depth. This action takes place virtually instantaneously, whereas comparative tests with chlorine have shown high doses and long contact times to be necessary. The amount of ozone necessary (to reduce colors of 50-60° Hazen to below 5° Hazen) would normally be between 1.5 and 4 mg/liter. High concentrations of ozone in air are desirable, and two-stage injection is sometimes required.

Microzon and MD Processes†*

Many surface waters are relatively clear and can be satisfactorily clarified by *Microstraining;[8] others are often colored, and, for this reason alone, are treated by chemical coagulation, sedimentation, and rapid sand filtration. As the substances producing color are normally derived from the soil and surface vegetation of the impounding area, they are usually of organic nature and can be removed or reduced by ozonation. The combination of Microstraining and ozonation (Microzon) was first installed in 1962 at Baerum Waterworks, Oslo, Norway, with a flow capacity of 19.5 mgd. In 1967, a larger project was commissioned at the Loch Turret Waterworks, Scotland, with flow capacity of 27 mgd. The consulting engineer of this scheme has stated that the capital costs are about half the corresponding figure for conventional treatment and operating costs about one quarter of those which would have been involved in the purchase of chemicals for conventional treatment. A point of great importance is that no chemical sludges are produced.[9] Other Microzon investigations are under way in Europe, including projects for Loch Lomond, Scotland; and Manchester, England.

---

*Registered Trade Mark
†Patented in U.S.A.

In cases where the raw water carries organic colloidal turbidity presenting well known difficulties in coagulation by conventional treatment, it has been found that ozonation can effect fundamental changes in the nature of the colloids and make them susceptible to coagulation with minimum chemical doses producing microflocs removable by high rate rapid sand filters without the need for sedimentation. This process of colloidal modification is called Micellization/Demicellization (the MD process). The sequence of treatment is as follows.

1. Mechanical filtration by Microstraining to remove suspended solids.
2. Oxidation of the organic colloids by ozonation (the Micellization stage).
3. Coagulation using a reduced chemical dose to produce microflocs (the Demicellization stage).
4. High rate rapid sand filtration for the removal of microflocs, employing the sand bed in depth. The original microscopic suspended solids having been removed no "schmutzdecke" is formed on the surface of the sand bed thus permitting high filtration speeds.
5. Final treatment with a sterilizing agent, which can be ozone, taken from the Micellization stage.

A simplified explanation of Micellization/Demicellization is that ozone, by oxidizing the hydrophilic organic colloids which are difficult to coagulate, effectively changes them into hydrophobic forms of inorganic nature which are readily coagulated. Installations of this type are in operation at Roanne, France (9.25 mgd), and Constance, Germany (16 mgd). In Canada, tests are currently in progress at Winnipeg and preliminary results have been reviewed.[10]

MUNICIPAL WASTEWATER TREATMENT

In the treatment of municipal wastewaters, ozone is normally applied following the primary and secondary sewage treatment processes. It is advantageous to use a filtration stage, such as Microstraining, for tertiary treatment where possible. This means that the water would be virtually free of suspended matter, and ozone would be able to act on the dissolved organics and other oxidizable substances. The effect of ozone in the treatment of wastewaters is similar to its effect on municipal water supplies, as described above.

Ozone will kill bacteria and viruses more quickly and thoroughly than chlorine, and with lower doses. It also reduces BOD and COD and, using higher doses, can oxidize refractory organics. The absence of a persistent residual is another advantage of ozone because contaminants are removed without producing secondary pollutants and without increasing the inorganic salt concentration.

A number of tests have been made in this country and abroad on ozonation of sewage effluents.

In Redbridge, London, tests were carried out using the MD process on a trickling filter effluent, following final settlement. Equipment comprised Microstraining with Mark 'O' fabric, followed by two ozone contact columns in series in which ozone doses up to 25 mg/liter could be applied to a flow of 40 gpm. Terminal sand filtration was operated at rates up to 4 gpm/sq ft. Test results showed that Microstraining reduced suspended solids by an average of 61%. The most obvious effect of ozonation was to remove the color of the effluent, giving a product which, after sand filtration, was clear and sparkling, comparable in appearance with potable water. After ozonation, the effluent was fully saturated with respect to dissolved oxygen. Removal of detergent residue by ozone was appreciable, the anionic material being reduced from 1.1 to 0.2 mg/liter and the non-ionic from 0.3 to 0.05 mg/liter. Some of this removal appeared to be due to foaming in the ozone columns.

Microbiological examinations showed that Microstraining alone had little effect. Ozone killed the vast majority of organisms present, including all the Salmonella and viruses. Sand filtration made little further improvement.

Opening a discussion on this work in England recently, G. A. Truesdale of the Water Pollution Research Laboratory summarized the results obtained[11] by saying that ozone had produced an effluent which, from a microbiological point of view, was far superior to river waters normally used as sources for domestic supplies and which would be suitable for a variety of industrial purposes without further treatment. In some cases, treatment by Microstraining and ozone alone may be sufficient to permit water reuse.

At Totowa and Berkeley Heights, New Jersey, trickling filter effluent following chemical coagulation and settlement was treated by ozone.[12] Virtually all color, odor, and turbidity were removed, and COD was reduced below 15 mg/liter. Bacteriological tests showed that no live organisms remained, and surface active detergents were removed. Ozone concentrations from 11 to 48 mg/liter, in oxygen, proved equally effective.

In these tests, various methods of dosing the water
with ozone were tried; including passing the water downward through a packed tower through which ozone was admitted in an upward direction. Next, using a batch reactor,
the water was mixed by a high-speed turbine agitator, with
ozone gas introduced adjacent to the blades. Finally, a
gas injector system similar to the Otto emulsion column
was used.

From the results obtained, a theoretical model of a
six-stage contactor using gas injectors in each stage was
developed, water passing through the columns in series
and ozone gas being admitted to the injectors in parallel.
By this method, a high dosage of ozone can be effectively
applied without loss of efficiency by decomposition. Since
oxygen is used for production of ozone, it must be recycled
from the last stage and recovered.

Following these initial tests, a six-stage reactor was
installed at the Blue Plains Sewage Treatment Plant in
Washington, D.C. Ozone tests were carried out on effluent
from activated sludge processes after tertiary treatment.
The effluent from a physical/chemical sewage treatment
plant was also evaluated. Flow capacity of the ozone
system was approximately 35 gpm. Initial results indicate
that the values obtained in the feasibility tests can be
repeated on a larger scale. Further details are contained
in a report presented elsewhere in this publication.

At Hanover Park in Chicago, Illinois, tests using
ozone were made on activated sludge effluent following
tertiary treatment by Microstraining. The results showed
that an ozone dose of 6 mg/liter was effective in reducing
bacterial counts below 2000 per 100 ml (fecal coliform)
and 5000 per 100 ml (total coliform). BOD reductions
averaging 30% were also obtained, and reductions were
noted in color, turbidity, cyanide, and phenol.

Ozone was applied in a two-stage Otto Reactor to a
flow rate of 91 gpm, contact time in each stage being
approximately 5 minutes. The uptake of ozone was very
rapid, and it was found that only one stage was necessary
for effective treatment. Suspended solids in the Microstrainer effluent ranged from 1 to 7 mg/liter and BOD
from 4 to 11 mg/liter.

In Philadelphia, tests have been carried out on ozonation of storm water overflows from combined sewers. This
pollution problem is extensive in the larger and older
cities. At onset of a storm, water is abstracted from a
receiving sump and passed through a Microstrainer before
being treated by either chlorine or ozone. Because of the

difficulty of running tests during a storm period, the
Microstrainer effluent is collected in a large storage
tank from where samples are taken for subsequent chemical
testing.

Ozone is applied in a two-stage Otto Reactor to a flow
rate of approximately 20 gpm. Facilities are provided to
use up to 50 mg/liter ozone although the maximum dose used
so far has not exceeded 10 mg/liter. Initial results[13]
from the chlorine and ozone tests were inconclusive,
although indicating that substantial disinfection, up to
98% removal of bacteria, could be obtained with doses
between 5 and 10 ppm.

INDUSTRIAL WATER TREATMENT

Ozone is used for disinfection and destroying organic
matter particularly where the absence of a permanent
residual is important, as in the pharmaceutical industry.
In the food industry, the beneficial effects of ozone include the destruction of tastes and odors which may be
present in the water supply. In the manufacture of butter,
for example, water-borne bacteria and organisms can cause
black spots and give the final product a slimy feeling.
The use of ozone will eliminate these problems.

In the beverage industry where flavors are developed
from organic compounds, the presence in the water supply
of organic matter will create objectionable reactions,
destroying or modifying the flavors. Chlorine is unsuit-
able for treatment of these waters, since the permanent
residual would create taste problems. The reaction of
chlorine with compounds in the water may also accentuate
some tastes. Ozone is, therefore, employed. Ozonated
water is also suitable for producing artificial ice,
because it eliminates the odor and taste of chlorine or
hypochlorite, phenols, and hydrogen sulfide. If waters
containing these impurities are boiled for cooking or
steaming foods, the food flaver can be seriously affected.
Ozone is beneficial, since it destroys the organic matter
without leaving unpleasant residuals.

Ozone is also used for sterilizing industrial containers, such as plastic bottles where heat processes are not
applicable and chlorine compounds would cause destruction
of the plastic material. The short-term residual of ozone
is a useful factor in ensuring sterilization of the con-
tainer, and, since there is no permanent residual, the
taste of the beverage is not affected by the bottling.

Breweries use ozone as an antiseptic and selective agent in the work of yeast and in fermentation. Ozone acts on the yeast before fermentation of yeast seeds by destroying the pathogenic ferments without harming the yeast itself. Ozone is also applied to the controlled sterilization of nutritive solutions used in yeast production.

Ozone is used in the treatment of swimming pools, for the treatment of water used in medical therapy, and for water treatment in zoo aquariums.

The application of ozone for sterilizing sea water used in the purification and washing of shellfish has been common for many years. Excellent shellfish vitality is maintained, thanks to the oxygenation effect. There is no change in the mineralization of the sea water and no harmful effect on marine plankton.

Slimes created in cooling towers can be controlled by ozone. It may be introduced into the holding tank underneath the tower or into the water as it flows over the trays. Ozone will destroy microorganisms, algae, and slime-forming bacteria.

Ozone can be used for destroying organics upstream of ion exchange. In preliminary laboratory tests using an artificially contaminated tap water, ozone was found to increase by 30% the amount of water which could be treated up to breakthrough of conductivity. Further tests are proceeding to evaluate the best method of contact and release of ozone. This could be an important application, since increasing pollution of surface waters has caused difficulties in the operation of ion exchange systems.

In all these applications for industrial water treatment, the dosage level of ozone is relatively low, normally below 10 mg/liter and most often not exceeding 1 or 2 mg/liter.

INDUSTRIAL WASTEWATER TREATMENT

Many industrial wastes contain organics which are decomposed by biological processes similar to those used in sewage treatment. Ozone can be used subsequently to further purify the wastewater by destroying refractory organics, reducing BOD and COD and disinfection.

Effluents from industrial processes containing phenols[14] can be effectively treated by ozone. At an oil refinery in Canada where the final effluent standard is less than 0.015 ppm phenols, biological purification followed by ozone is employed. Biological purification is achieved by a

combination of the activated sludge process followed by
trickling filters. The effluent is then settled before
being pumped through a large stainless steel contactor,
20 feet high and 15 feet in diameter, where ozone is
introduced through a diffusor. Reduction of phenol in
the final effluent takes place from 0.38 ppm down to
0.012 ppm (96% reduction). The total ozone production
is approximately 190 lbs per day, produced from air, and
flow capacity of the treatment process varies between 300
and 600 gpm. The dose of ozone is between 20 and 40 ppm.

Ozone has also been tested for the treatment of wastes
from a synthetic polymer plant.[15] This effluent contained
biorefractory organics which were attacked and largely
destroyed by ozone. Biological processes could not be
used for purification because the effluent contained
chemicals toxic to microorganisms.

Various effluent streams were tested in a 100-ml
batch-type reactor using ozone generated from oxygen.
All the streams contained some unsaturated organics, and
one also contained sodium 8-alkyl napthaline sulfonate-2
(SANS), which is very resistant to biodegradation. The
COD of the wastewaters was reduced significantly by ozonation when the oxidizable organics consisted of unsaturated
compounds. Ozonation of the SANS gave partial destruction
but apparently did not render the compound more biodegradable. Very high doses of ozone were employed, up to
6,000 ppm, at which level COD removal approximated 90%.

At a large industrial plant in Kansas, more than 350
lbs of ozone is used each day for secondary wastewater
treatment to deal with effluents containing cyanide,
phenols, oil, detergents, sulfides, and sulfites.[16] The
dose of ozone is approximately 20 ppm, applied in two
diffusor contact towers through which the water passes in
series. Ozone dosage is preceded by primary treatment
consisting of oil skimming, sulfur dioxide treatment of
chromates, pH control, coagulation, and clarification.
The final effluent after ozonation passes into a lagoon,
where the absence of toxic wastes and pollution is shown
by the presence of live fish.

The economics of using ozone for the treatment of acid
mine drainage has been evaluated in a preliminary study
for the Environmental Protection Agency.[17] Treatment
would consist of oxidation of ferrous to ferric iron with
ozone, and neutralization with limestone. The report
concludes that oxidation by ozone compares favorably with
the present air-oxidation system. Conventional treatment
methods are estimated to range from $0.10 to $1.30/1000

gallons, whereas the ozone process costs from $.09 to
$.78/1000 gallons. The study also mentions that the
ozone-limestone system offers the potential of simplified
process control, higher plant throughput, removal of
additional pollutants from acid mine drainage, and reduced sludge handling requirements.

Tests carried out in France on the effluent from a
large metal finishing factory showed that an ozone dose
of 80 to 90 mg/liter could remove 25 mg/liter of cyanide.[18]
The results of initial pilot tests are as follows:

Cyanide content of effluent before ozonation =
25 mg/liter; cyanide content of effluent after
ozonation = 0.

Concentration of ozone in air (gr/cu m)	7	14	20
Total ozone applied (lbs/per lb cyanide)	7.3	5.7	4
Ozone lost to atmosphere (lbs/per lb cyanide)	3.8	2.5	0.8
Ozone used in destruction of cyanide (lbs/per lb cyanide)	3.5	3.2	3.2

These results show that total destruction of cyanide with
ozone is easily obtained. The process is not only tech-
nically feasible, but also economical compared with other
processes.

The specific dose of ozone diminished when the concentration of cyanide increased and when the concentration of
ozone in air increased. The loss of ozone from the contact
columns was reduced by increasing the concentration of the
ozonized air.

In full-scale installations designed as a result of
these pilot tests, a flow of 20 cu meters/hr (88 gpm) is
already being treated. A further installation treating
another 20 cu meters/hr is projected. For each project,
four Ozonators, each containing 40 vertical plate elements
and specially designed to give high concentrations of ozone
produced from atmospheric air are used. The ozone is injected into the effluent in diffusor tanks with a contact
time of approximately 25 minutes.

For each project the maximum ozone production is 1250
kg/hr (267 lbs/hr) at a concentration of 12.5 gm/cu meter
and 650 kg/hr (134 lbs/hr) at a concentration of 30 gm/cu
meter.

In the treatment of industrial wastes, it is generally assumed that 2 parts of ozone will be required for the destruction of 1 part of phenol and 1-1/2 parts of ozone for each part of cyanide. However, because other chemical compounds in the effluent could materially alter these requirements, it is generally desirable to carry out laboratory tests in all cases.

REPRESENTATIVE COSTS OF OZONATION

Installation Costs

Since ozone cannot be bottled, its capital cost will always be higher than that of chlorine. In many cases, this higher cost will be justified because ozone can make a contribution that chlorine cannot. Also, in the Microzon and the MD processes, the cost of ozone should be compared not only with chlorine but also with the clarification equipment and chemical treatment superseded.

Installation costs increase with the applied dosage, and the graph shown in Figure 51 is prepared on the conventional basis of 1 ppm of ozone required for water sterilization. The costs include ozone generators and piping, air preparation equipment, and electrical equipment, but no construction work.

The following figures were prepared for the Loch Turret Microzon project based on a flow of 27 mgd and ozone dosage of 2.5 ppm.

Costs of Treatment System, Loch Turret, Scotland:	
Main building, treatment chambers channels, air and water piping, electric cabling, instrumentation, and other furnishings	$ 714,000
Microstrainers	240,800
Ozonizers, related transformers and associated air preparation and injection equipment	400,400
Chlorinators, circulating pumps and other auxiliary equipment	14,000
Access roads, transmission lines, overflow works, and other external services	67,200
Total	$1,436,400

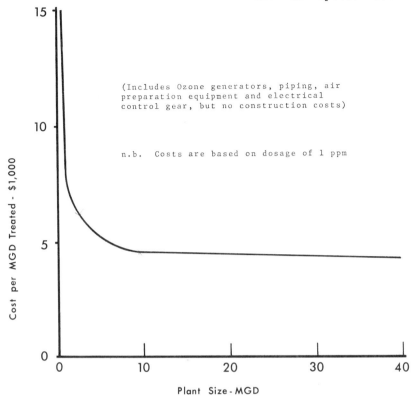

(Includes Ozone generators, piping, air preparation equipment and electrical control gear, but no construction costs)

n.b. Costs are based on dosage of 1 ppm

Plant Size - MGD

Figure 51. Installation Costs of Ozonation Equipment

Operating Costs

Operating costs for ozonation chiefly depend upon the cost of electricity. Because ozonation would normally be added on to other treatment processes, the tariff charge is usually at the low end of the scale. Assuming a cost of 1 cent per unit (kwh) and an electrical consumption of 28 watt hours/gram of ozone, including ozone production and injection, air preparation and ancillaries, the cost of treating 1000 gallons at a dose of 1 ppm would be 0.1¢. In comparing this figure with that of chlorination, not only would the cost of the chemical need to be taken into account but also the cost of transporting and storing chlorine containers.

Electrical Loads, Loch Turret Treatment System:

Equipment	Electrical Load (kw)	Percentage of Total
Microstrainers	18.7	5.0
Air preparation	27.7	7.5
Ozone production	287.2	76.0
Ozone injection	43.6	11.5
Total	377.2	100.0

Total Costs

If the capital cost of ozone equipment for treating 10 mgd is amortized over a period of 20 years at 6% interest plus taxes, this charge would amount to approximately 0.2¢ per 1000 gallons. Therefore, an average figure for installation and operating costs, based on a dosage of 1 ppm ozone, would be 0.3¢ per 1000 gallons treated.

OTHER APPLICATIONS OF OZONE

The principal usage of ozone in the United States is in the chemical industry. Ozone is used as an oxidizing agent in the production of oleic acid in the manufacture of plastics. Ozone is also used in making carbon black for high quality inks, and for the production of peroxyacetic acid.

Another important nonwater application of ozone is in the control of odors from sewage treatment plants, manufacturing facilities, hospitals, kitchens, etc. In these cases, ozone is applied to the exhaust air, given a short period of contact, and then discharged to atmosphere. The oxidizing reaction takes place very quickly and is effective in destroying organic odors. Representative doses would be between 1 and 2 ppm by volume, with a contact time of a few seconds to half a minute.

There is development work proceeding in the paper and clay industry to evaluate the use of ozone as a bleaching agent. Small quantities of ozone are used in the manufacture of drugs. Ozone is also used for preserving foods in storage.

SAFETY ASPECTS

No paper on ozone would be complete without a reference
to the safety aspects of using ozone, which is a toxic gas.
Numerous reports on air pollution have stressed the formation of ozone in the atmosphere as a result of photochemical
reactions and have indicated the long-term danger to plants
and animals. As a result of this publicity, some engineers
believe that ozone is a dangerous chemical to use.

The reverse is true. Ozone is, in fact, far less
hazardous in water and waste treatment than the gaseous
chlorine which is extensively employed. This is because
the ozone is generated and used only in low concentrations
and is not stored under pressure. Any escape of ozone
from a treatment system can be quickly stopped by turning
off the electric supply.

Ozone is self-policing. In concentrations far below
harmful or toxic levels, it is immediately noticeable by
irration it produces in the nasal passages. Concentrations
up to 20 or 30 times higher than this and prolonged exposure over many hours are required before the gas can
be harmful (Figure 52). Furthermore, all the equipment
used to generate ozone is protected by fail-safe devices
as described above. Most ozone is generated at atmospheric
pressure or at pressures not exceeding 8 to 15 psi, so
that any leakage is of relatively small proportions.
Chlorine, however, is normally stored under pressures
which produce liquefaction. If a leak develops in the
chlorine containers, they are difficult and dangerous to
seal because of the rapid escape of the gas. Sodium
hypochlorite is now used in several large cities for
water disinfection to avoid the hazard of handling and
storing gaseous chlorine.

SUMMARY AND CONCLUSIONS

Ozone is a powerful oxidizing agent, about twice as
powerful as chlorine, used for sterilizing water supplies
in Europe and Canada. Its use is not common in the United
States in the treatment of potable water, principally because of the absence of a residual in the distribution
system. This may, however, represent an advantage in the
treatment of waste treatment plant effluents, since there
will be no toxic effect in the receiving stream, eliminating
secondary pollutants, and the build-up of inorganic salts.

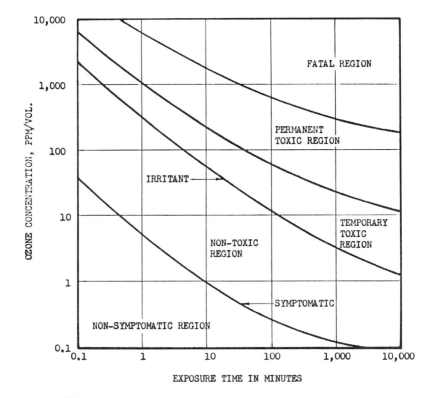

Figure 52. Ozone Toxicity

Ozone cannot be bottled and must be generated on site, which avoids the hazard present in carrying chlorine containers through municipal areas. In comparative tests with chlorine, ozone has been found more effective in killing bacteria and viruses and in removing organic color, taste, and odor. Its action is virtually instantaneous, requiring no large contact tanks.

In the treatment of industrial effluents, ozone is used to reduce BOD and COD and to destroy objectionable chemical compounds such as phenols and cyanides.

The combination of Microstraining and Ozonation (Microzon) has been tested on municipal sewage effluents in London and Chicago. Microstraining will remove the majority of suspended solids contained in the secondary effluent, and ozone will kill the vast majority of organisms present, including all bacteria and viruses. A

noticeable effect of ozone is to remove the color of the
effluent. Other beneficial effects include high dissolved
oxygen and removal of detergent residues.

When applying relatively large doses of ozone, special
systems must be used for contacting the water. Normally,
ozone is applied in parallel to several stages through
which the water flows in series. By this means, the water
receives successive increments of ozone which is applied
as soon as possible after generation to prevent any reduction in its oxidizing power. A multicontactor system of
this type is at present under investigation on sewage
effluents in the Blue Plains Plant in Washington, D.C.

In Philadelphia, a Microzon system is being tested on
storm water overflows from combined sewers.

Modern ozone generators employ electrode systems made
up of a series of tubes or plates equipped with insulators
and provided with cooling arrangements. Specially designed
transformers step up the normal mains voltage to 15,000
or 20,000 volts. Air preparation equipment comprises
filters and driers using refrigeration and desiccation.
The design of these units has been well established over
many years, and they incorporate high quality materials
completely protected by fail-safe devices. Ozone costs
approximately 8 cents per pound produced from air and 3.5
cents per pound when made from oxygen. Installation costs
range from $500 to $1,000 for each pound per day of ozone
generating capacity.

An outstanding advantage of ozone is its immediate
and effective attack on viruses in water supplies.
Medical evidence shows that dilutions of polio virus can
be rendered inactive in 2 minutes by a small dosage of
ozone, whereas chlorine may take several hours and much
higher doses. This aspect of ozone will undoubtedly
bring it into more common use in this country as the
search continues for higher quality water supplies.

ACKNOWLEDGMENTS

Acknowledgment is made to the authorities and consulting engineers concerned for permission to refer to
the installations described in this paper. Thanks are
also due to Crane Co. and Trailigaz (France) for permission to publish details of their processes. Much of the
information on Ozonator design was contributed by M.
Guillerd, Director of La Compagnie des Eaux et de l'Ozone,
and published in *Techniques et Sciences Munipales*.

REFERENCES

1. Bringman, G., "Determination of the Lethal Activity of Chlorine and Ozone on *E. coli*." Z. Hyg. Infektionskronkh., *139*, 130 and 333 (1954).
2. Ingols, R. S., and Fetner, R. H., "Ozone for Use in Water Treatment." Proc. Soc. Water Treatment Exam., *6*, 1, 8 (1957).
3. Kessel, J. F., et al., "Comparison of Chlorine and Ozone as Virucidal Agents of Poliomyelitis Virus." Proc. Soc. Exptl. Biol. Med., *53*, 71 (1943).
4. Smith, W. W., and Bodkin, R. E., "The Influence of Hydrogen Ion Concentration on the Bactericidal Action of Ozone and Chlorine." J. Bacteriol., *47*, 445 (1944).
5. Ridenour, G. M., and Ingols, R. S., "Inactivation of Poliomyelitis Virus by 'Free' Chlorine." Amer. J. Public Health, *36*, 639 (1946).
6. Whitson, M. T. B., "The Use of O3 in Purification of Water." J. Roy. Sanit. Inst., *68*, 5, 448 (1948).
7. Ozonation—Polish Report to the International Water Supply Congress and Exhibition, Stockholm, 1964.
8. Diaper, E. W. J., "Tertiary Treatment by Microstraining." Water and Sewage Works, *116*, 6, 202 (1969).
9. Campbell, R. M., "The Use of Ozone in the Treatment of Loch Turret Water." J. of Inst. of Water Engrs., *17*, 4, 333 (1963).
10. Sommerville, R. C., and Rempel, G., "Ozone for Supplementary Water Treatment." Presented at the 91st Annual Conference of the Amer. Wat. Wks. Assoc., Denver, Colorado, June 1971.
11. Boucher, P. L., "Microstraining and Ozonation of Water and Wastewater." Proceedings of the 22nd Indus. Waste Conf., Purdue University Engr. Ext. Series.
12. Huibers, D. Th. A., McNabney, R. L. and Halfon, A., "Ozone Treatment of Secondary Effluents from Wastewater Treatment Plants." Report No. TWRC-4, Adv. Waste Treat. Res. Lab. and the Air Reduction Company (April 1969).
13. Cochrane Division, Crane Co., "Microstraining and Disinfection of Combined Sewer Overflows." Report No. 11023 EVO, Adv. Waste Treat. Res. Lab (June 1970).
14. ———, "Phenol Free Waste Water." Chemical Engineering, *66*, 8 (1959).
15. Kwie, W. W., "Ozone Treats Waste Streams from Polymer Plant." Water and Sewage Works, *116*, 2, 74 (1969).

16. _____, "Ozone Counters Waste Cyanides Lethal Punch." Chemical Engineering, *65*, 3 (1958).
17. Brookhaven National Laboratory, "Treatment of Acid Mine Drainage by Ozone Oxidation." Environmental Protection Agency, WQO, Water Pollution Control Research Series, Report No. 14010 FMH 12/70 (1970).
18. Guillerd, J., and Valin, C., "Traitement par l'Ozone." L'Eau, *48*, 5, 138 (1961).

INDEX

183